THE
future
IS
now

CONTROL OF NATURE

Series Editors

Morton L. Schagrin
State University of New York at Fredonia

Michael Ruse
Florida State University

Robert Hollinger
Iowa State University

THE
future
IS
now

SCIENCE AND TECHNOLOGY
POLICY IN AMERICA
SINCE 1950

ALAN I MARCUS
& AMY SUE BIX

Humanity
Books

an imprint of Prometheus Books
59 John Glenn Drive, Amherst, New York 14228-2119

Published 2007 by Humanity Books, an imprint of Prometheus Books

Inquiries should be addressed to

Humanity Books

59 John Glenn Drive

Amherst, New York 14228-2119

VOICE: 716-691-0133, ext. 210

FAX: 716-691-0137

WWW.PROMETHEUSBOOKS.COM

11 10   5 4

Library of Congress Cataloging-in-Publication Data

Marcus, Alan I, 1949–
    The future is now : science and technology policy in America since 1950 / by
Alan I Marcus and Amy Sue Bix.
        p. cm.
    Includes bibliographical references and index.
    ISBN 978-1-59102-472-9 (pbk. : alk. paper)
    1. Science and state—United States—History—20th century. 2. Technology
and state—United States—History—20th century. I. Bix, Amy Sue. II. Title.

Q127.U6M278 2006
338.97306—dc22

                                              2006027772

CONTENTS

INTRODUCTION

Science and technology are the essential building blocks of contemporary American society and governance. "Smart" bombs and other sophisticated weapons replace countless soldiers. Hybrid seeds, biotechnology, and fertilizers are the basis of agricultural productivity. Job creation, global competitiveness, and the American standard of living stand as the consequence of scientific insight and technological application. Acquisition of knowledge and provision of information and communication, whether through electronic impulses or on land, rest on science and technology.

The effects of science and technology are immediate and abundant. In a very real sense, what we do in the present dictates the shape of the future. Put another way, American society and government's dependence on science and technology means that the future is now.

The federal government has fully recognized science and technology's absolute centrality to contemporary existence. The greatest share of the federal budget goes to pay for services and products that are science- or technology-based. But governmental appreciation of science and technology as its most important and most expensive elements is not new. Shortly before 1950, many American political leaders and others began to acknowledge that America's World War II success stemmed directly from its scientific and technological prowess. They also acknowledged that continued success in both military and commercial spheres required nurturing. These sentiments manifested themselves most explicitly in the 1950 creation of the National Science Foundation (NSF), the first federal agency devoted entirely to the sponsorship and funding of scientific research.

Establishment of the NSF marked the beginning of the federal government's marriage to science and technology that has now spanned more than a half century. The romance has remained impressively unflagging, even as the intensity of the infatuation has deepened. Favoring and fostering science and technology has been federal policy for more than fifty years. Yet it would be wrong to suggest that continual, persistent support of and reliance upon science and technology means that the federal government's approach to nurturing and employing them has remained unchanged. Nothing could be further from the truth. Great differences have occurred and disputes have erupted over how to best achieve those ends. Americans have repeatedly disagreed over techniques, programs, methods, and interests as they endeavored to find the most successful means to advance that national policy. This was true even as policy was being enacted. Many times the nation rearranged, shifted, and reconceptualized its federal science and technology policy while its ambitions—furthering and using science and technology—remained constant. What needed doing was to analyze and explore the means, mechanisms, and methods by which various constituencies pressed, reshaped, and refashioned the federal bureaucracy in pursuit of the nation's ambitions.

Two works have begun to address the federal government's interaction with science. Both Bruce Smith's *American Science Policy Since World War II* (Brookings Institution, 1990) and Alexander Morin's *Science Policy and Politics* (Prentice-Hall, 1993) are program-specific, deal with the inner workings of a bureaucratic group or Congress, and are heavily dependent on viewing science as a subsidiary of national defense. Their message is clear. America needs an effective, clearly stated, apolitical, and unchanging science policy if it is to maximize efficiency.

Wonks and others find these books useful, but we have tried to do something more. We do not personify science, treat it as a nonpartisan actor among a cast of partisans. Nor do we ignore technology. Technology has an important place in the formulation. We attempt to employ science inclusively in our analysis. We discuss basic research, the applications of science and even social science—the manipulation of people and process as well as the understanding thereof—which rarely figures in science policy analyses.

Our broad, comprehensive outlook cannot and does not try to be a day-to-day, department-by-department rendering of science and technology policy. It uses instead scenes, case studies, or vignettes as indicative and illustrative of larger, important policies and programs. The studies are illustrative of the approaches, activities, and problems found in a wide variety of government initiatives. We also highlight what we see as critical and underappreciated areas, such as universities, which have been increasingly dependent on federal dollars and have increasingly become the engines through which national science and technology policy is revealed. We also look at the roles of gender and minorities and how they are embedded in federal programs, the application of social science to the processes of funding and selecting research agendas, and technological initiatives as a justifiable means to discover and apply behavioral techniques.

The volume is organized chronologically, which gives way in the 1960s to an organization based on presidencies. The presidency assumed center stage in the 1960s and later as presidents set the nation's agenda as part of their executive duties. Their administrative styles, their philosophical and political principles and ambitions, and their distinctive personalities all translate into what kind of subjects presidents favor, how they choose to interact with Congress, and what kind of relationships they wish to have with scientists and other acknowledged stakeholder groups. Rather than merely tracing the development of a particular policy—defense, for instance—over a series of administrations, this book focuses on how a presidential administration is central not only to overall science and technology policy but also to individual science- and technology-based initiatives. An analysis of the former type ignores the fact that science and technology policies and activities are located firmly in time, creatures of specific social, political, cultural, and scientific moments. It is those moments that help explain why actions are taken and rejected.

At the heart of any contemporary analysis of US science and technology policy is what is thought to be a rather simple question: What is or has been America's science and technology policy? Such a simple question is exactly the wrong question to ask, however. That contemporary Americans and others persist in asking it causes difficulties. By presuming that there has been—or should be—a discrete, definable, tangible policy, analysts put themselves into a box. When people expect analysts to be able to answer such a question, it hampers or prevents serious attempts to analyze the situation beyond it. If that mode of thinking were actually put into practice, the resulting policies would be unhealthy for America. This volume neatly demonstrates that the idea of a discrete, unified, consistent approach to science and technology over time needs to be revised.

It is possible to establish and mandate an unbending set of requirements and specifications for national policy, but to do so would court disaster. Situations are continually fluid, with huge unforeseeable economic and political dislocations and unknown relationships between economic policies, theories, and their goals. The past half century alone has proven that the problems and issues identified as confronting America change at such a prodigious rate that any attempt to set a precise course would certainly have to undergo revision after revision. Indeed, even if government claims to chart a "consistent" course today, that course will change virtually every six weeks.

US monetary policy is a good example. The Federal Reserve Board meets monthly to decide how monetary policy needs to be rearranged, and it generally changes the interest rates to stimulate the type of activity it seeks. But even that practice is less than precise.

Within the last couple of years, Federal Reserve Board members have noted that certain economic activity expected from interest-rate tinkering just has not happened; the idealized mathematical model upon which they have operated does not reflect the real world. Put baldly, when we say we have a consistent monetary policy, not only are we wrong, but that expression also obscures how American government truly functions.

By talking about a static policy, proponents of models build their wares for a world that does not exist. It is a world without change or, at best, with scheduled or predicted changes. Such model builders would lack the constancy of questions that need to be addressed or resolved and therefore the certainty of their approach to those questions would be compromised. Surely, an idealized world would make matters easy. You could anticipate and predict, then dictate a certain, specific course for the future. But by searching for that perfectly artificial world, you privilege that approach and thereby lose the absolute essence of what Americans actually do and of how incredibly well it works.

Democracy is the strength of American governance. And that is how policy is actually made. It is made through battles among constituencies in which the executive branch has by far the greatest say. Someone once said that policy is a lot like sausage; you may enjoy the taste but you won't want to see it being made. It is necessarily messy, inexact, dirty, and flawed. So too is modern existence. Policy in the real world does not conform to some sort of neat social science formulation. That is not to suggest that planning, prediction, and thought are not—or have not been—critical aspects of past science and technology policy making. They have been integral and valuable aspects in the past, but their true utility is as recommendation, not proscription.

In American representative democracy, there is no shortage of persons and professions to outline their systematic and ad hoc plans/analyses and demand acceptance/implementation. It is the genius of American politics that one and several people sift through all the advice and select a course to pursue, and that persons who are subsequently thrust into the role of choosing a course are rarely bound to their predecessors' course. Each electoral contest sends us in another direction; topics and subjects given short shrift in the past may suddenly capture center stage. In a democracy, there is often lurching back and forth, perhaps—intolerable inefficiencies to model builders—but excesses get smoothed out and consensus gets enhanced. In this way, new ideas and actions can enter the policy arena and be heard and considered without a commitment of permanence. Similarly, bad ideas can be removed and rejected in a relatively small space of time. No one group, including those people who claim it is their special expertise to set policy, can capture the policy apparatus; they are a voice, not the voice.

Chapter One

IN THE BEGINNING

FROM MANHATTAN PROJECT TO NATIONAL SCIENCE FOUNDATION

The American World War II effort depended on new and improved weaponry: proximity fuses to detonate bombs shortly before they reached ground to maximize explosive power and damage; jet engines to replace the heavier, less powerful, and more complicated piston engines; radar to spot enemy forces prior to detection by more conventional means; and penicillin to fight battlefield infections. Each of these weapons resulted from a single rationale: teams of scientists and engineers were directed to fashion specific devices to enhance allied military capabilities. Creation and use of the atomic bomb stemmed from similar circumstance, albeit on a far grander scale. But even before the bomb was constructed, President Franklin D. Roosevelt had learned enough about the awesome power of science to decide it should become an essential part of American life.

Ironically, Roosevelt did not focus on science's military prowess. He wondered instead of its ability to shape the future American economy. As part of his New Deal to combat the Great Depression, Roosevelt had established numerous federal organizations and agencies to enable Americans to fulfill what in the 1930s commonly became known as "the American Dream" and to participate in the "American Way of Life." The New Deal incorporated or was derived from sentiments not inconsistent with Keynesian economics. Demand, not supply, was the key economic discriminant. America's economy would rebound and blossom only when Americans had funds to purchase commodities and the demand for those commodities would encourage manufacturers to increase supply, which would lead to increased employment. In this economic view, demand was plastic and could be created or manipulated, and therefore many New Deal programs worked either to get money into citizens' hands to purchase new material goods or to advertise as desirable a new material possession.

Roosevelt saw science as a means to create jobs and, therefore, demand. He asked Vannevar Bush, director of the Office of Scientific Research and Development—the agency in charge of science and technology in service of the war effort—to examine how the fruits of wartime science "in the days of peace ahead" might be transferred to American industry. This science, Roosevelt concluded, can "stimulate new enterprises, provide jobs for our returning servicemen and other workers, and make possible great strides for the improvement of the national well-being." As critical, Roosevelt also sought Bush's advice about how the federal government could ensure a continual flow of new scientific insight and, therefore, jobs, "improvement of the national health," and the "betterment of the national standard of living" for subsequent generations.[1]

FDR died before Bush finished his report on how science would fulfill the American dream and continue the American way of life. His report, delivered to President Harry S. Truman on July 25, 1945, set as axiomatic three things. First, science was fundamental to American society and life "because without scientific progress no amount of achievement in other directions can insure [sic] our

health, prosperity, and security as a nation in the modern world."
Second, scientists could not be isolated nor their work classified or
restricted because science and scientists "can be effective in the
national welfare only as a member of a team." Third, sources outside
the scientific community could not direct scientific inquiry because
"essential, new knowledge can be obtained only through basic scientific research," research stemming from the curiosity of scientists.
Bush found little in his contemporary situation that might support
his axioms. He noted that philanthropic contributions for scientific
research in America were diminishing and that war-ravished Europe
now lacked the facilities and funding to generate vast amounts of
new science. To guarantee that America would remain a vital nation,
Bush demanded that the federal government step into the science
breach. Taking an active, leading role in scientific inquiry hardly
constituted a radically new direction for American government,
Bush argued, because since the Lewis and Clark Expedition in the
early nineteenth century, the federal government had fostered "the
opening of new frontiers." The "modern way to do it" was by "pro-
moting scientific knowledge and the development of scientific talent
in our youth."

Bush envisioned government as the primary financier of
research, and government and industry as partners in the application of research. Colleges and universities would concentrate "their
research efforts to expanding the frontiers of knowledge," not to
applying "existing knowledge to practical problems," since those
institutions felt the "least pressure for immediate, tangible results."
Industry and government focused on immediate applications and
"use of public funds" would fuel the two prongs of the enterprise. To
ensure its success, the plan required the regular, consistent production of more and more scientists, and Bush again targeted colleges
and universities as the suitable educational institutions. But he also
sought to enlarge the pool from which scientists emerged. "Ability,
and not the circumstance of family fortune" ought to dictate who
became a scientist. Therefore Bush believed it a governmental
responsibility—a national duty, in fact, since scientists controlled
America's fate—to fund students who otherwise were unable to

pursue scientific or technical careers, or jobs that would prepare them for a future leadership role in the new scientific democracy. A scientific advisory board would coordinate these new scientific activities and ensure as much as possible the free flow of scientific information without compromising national security.[2]

Almost immediately after Bush delivered his report, the intellectual landscape changed. America detonated atomic bombs on Hiroshima on August 9 and Nagasaki on August 14, 1945. The awesome destructive force of these devices mesmerized the public and set off an important public conversation. Humbled by the bomb's sheer power, various religious organizations called for the nation to renew its commitment to moral precepts. A relatively small contingent of scientists who had worked on the bomb recognized the horror of their creation and formed the Federation of Atomic Scientists to vest control of nuclear power in civilian hands. Their angst was real and not just because they built the bomb; they had also urgently lobbied for it. Prominent nuclear scientists had advised President Franklin D. Roosevelt in 1941 that an investigation into a fundamental principle of science—conversion of matter into energy—could be transformed into a weapon of tremendous power, and they persistently argued that America must develop such an apparatus.

The bomb's moral ambiguities were merely one lesson from the war. More crucial and less controversial was the broad-based, widespread, and public agreement among Americans that science had won the war and would direct and determine the future. Again, the atomic bomb served as focus. Japan's unconditional surrender on August 14, 1945, without invasion of the island nation followed almost immediately from dropping the Hiroshima and Nagasaki bombs. (See fig. 1.) And if science was that important to the present and the future, was it not the responsibility of the federal government to foster and develop it? After all, the preamble to the US Constitution set the national government's responsibilities "to provide for the common defense" and "promote the general welfare" so as to "secure the blessings of liberty to ourselves and our posterity." In the aftermath of World War II, science seemed indispensable to the

http://en.wikipedia.org/wiki/Image:Hiroshima_aftermath.jpg

FIG. 1

The city of Hiroshima, Japan, after the American B-29 *Enola Gay* dropped the atom bomb nicknamed "Little Boy" on August 6, 1945. The force of the blast obliterated most buildings within a one-and-a-half-mile radius of the explosion, stripped trees bare, and literally melted the skin off human beings. The explosion and the effects of radiation exposure killed an estimated 70,000 to 80,000 people within a few months and at least as many over subsequent years.

nation's defense and military security, and the conventional wisdom was that the federal government must employ it to those ends.[3]

To be sure, the federal government had long sponsored science for military and defensive purposes. It justified scientific surveys as a fundamental facet of the nation's defense, even surveying inland waterways in the early nineteenth century for that purpose. During the Civil War, Lincoln established the National Academy of Sciences to aid the war effort. The Naval Observatory played an important scientific function during World War I.

Defense wasn't the sole recipient of governmental largesse. From 1887 the United States Department of Agriculture financed investigations into plant growth parameters and in 1925 expanded its purview to include social science investigations into the "rural home and rural life." The 1937 National Cancer Institute Act provided federal funding for cancer research, followed by other national institutes in the next decade, formed to provide moneys for scientific investigation of other diseases.

In the wake of World War II, government-sponsored research increased exponentially and it moved toward the defense-related sciences: physics and chemistry. Gaining the defection of German scientists, especially rocket scientists, became a national priority. Institutionalization of the Office of Naval Research as a permanent part of the military nexus set off unprecedented funding for research projects that seemed to hold naval or military potential. The office did not simply hire scientists and create numerous laboratories. It also depended on industry—private enterprise—and on universities and the scientists located at these schools. In both industry and academics, contracts were awarded by the office to investigate targeted questions.

What the office did was well in keeping with established federal precedent. Each initiative identified a defense-related problem and then aimed to design a precise scientific solution. But the most successful illustration of this procedure's efficacy—the atomic bomb—had also pointed to a different direction—that is, to nature. The creation of the atomic bomb had demonstrated that scientific research into fundamental questions of matter and life could result in solutions to problems. Yet attempts to direct or predict what possibilities might occur were impossible, even wrongheaded. In this view, knowledge—investigations into the nature of nature—was unfailing power but not necessarily immediate, clear-cut, predictable, nor useful.

An intriguing situation presented itself. Systematic investigation into natural phenomena, previously outside regular government purview, would yield solutions to America's problems—perhaps even to humanity's problems—but it was unknown which problems might be resolved and in what time frame. The future, after all, required precision planning and would be improved because of that planning, but nondirected investigations into natural phenomena rarely resulted in predicted outcomes; investigators could not guarantee exactly how the future would be enhanced and at what speed.

Problem- or situation-directed scientific research, and research agendas set by scientists, emerged as articles of faith after World War II, placing a premium on establishing the correct parameters to pursue the requisite tasks. The science-directed policy that would direct the future, and upon which Americans relied, was straightforward. But how that policy should be implemented remained a subject of considerable conjecture and debate. What was needed and what was necessary? Who dictated what and to whom? Would notions of efficiency prove more critical than notions of public accountability?

These basic questions of governance in a democracy shaped the new policy. Congress demanded accountability and oversight. The president insisted on control over any newly created executive department. Scientists demanded free and full disclosure of scientific information. Various sections of the federal bureaucracy recognized that if science controlled the future, unfettered access to scientific information might prove catastrophic. Scientists urged for science to be supported without reference to politics. Others maintained that the awarding of research funds to specific entities was little more than politics. What projects deserved what level of support? What should be the responsibility of those who received government grants? If the purpose was to direct the future, was it not in the nation's best interests to give the grants to the nation's best scientists?

None of these questions occurred in a vacuum, of course. American scientists contended that the past several decades had proved that political control of science had been disastrous and pointed to Nazi Germany and the Soviet Union as examples. Nazi scientists proposing various racial purity schemes consistent with the regime's theories of human difference found favor and research support at the expense of others, while in the Soviet Union Stalinist purges had driven scientists opposing communism out of their posts. Stalin's intervention also led the USSR to adopt a theory of genetics out of step with modern Mendelianism. Those in the Soviet Union refusing to embrace the approved Soviet brand of genetics were discredited, removed from their scientific posts, and denounced as reactionaries and enemies of the Soviet people.

In addition, post–World War II Europe was in tatters. It could no longer provide a bulwark against the Soviet Union, which, despite its alliance with Nazi Germany at the beginning of the war, emerged more powerful after the war. The Soviet Union's territorial ambitions were troublesome, and when Czechoslovakia and then much of the rest of Eastern Europe fell under its control, an "iron curtain of communism" enveloped much of the territory invaded earlier by Nazi Germany. The communist monolith expanded in 1949 when China, the world's most populous nation, fell to Mao Zedong and his communist forces. The Soviet Union's explosion of its first atomic bomb, also in 1949, fueled further fears of communism's world domination.

This terrifying communism-versus-capitalism world demanded immediate action. Americans relied on the tried-and-true method to counter immediate threats and situations. They targeted precise activities to resolve specific issues and questions and generally relied on the Department of Defense, the Department of Agriculture, the Department of Labor, and the executive branch. But in a more explicit attempt to ensure the nation's future, the federal government embarked upon a program to explore fundamental questions of science and to construct an infrastructure that would support scientific and technical inquiry and application, for both fundamental and more immediate issues, to combat the communist threat.

The newness of this initiative caused concern almost instantly. To be sure, Bush and his report stressing science as the harbinger of national prosperity and health had not disappeared. But the cold war climate of the later 1940s certainly colored the discussion. A sense of immediacy prevailed. General Dwight David Eisenhower, then president of Columbia University, wondered "what kind of government ahead." He urged Americans not to forget that government gained its powers from the consent of the governed and by refusing even "temporary surrender" of any freedoms. If they ignored his concerns, Eisenhower maintained, "the American Dream may become the American Nightmare." President Harry S. Truman looked outside the United States and argued that the fate of the world rested in American hands. Around the world, he said, the "communists are saying that they will bring food and clothing and health and a more secure life to . . . poverty-stricken people." To Truman, those people cared little about ideologies; they "will turn to democracy only if it seems to them to be the best way to meet their urgent needs."

"The benefits of freedom and democracy must be demonstrated to them," Truman said, urging Congress to "expand our programs giving technical assistance" to other nations. The purpose of this act, Truman noted, was "not to sell them automobiles and television sets," but rather "to help them grow more food, to obtain better education, and to be more healthy." Military men argued that Americans should "battle for the liberty of mankind" and be "aggressors for peace," while others wondered how higher education should function "in a period of armed truce."[4]

Despite the sense of immediacy and the intolerable cost of failure, Americans pursued the future rather optimistically. Pride in WWII success imbued the steadfast nation. The GI Bill had enabled millions of young men and women who had served in the armed forces during WWII to gain a college education at little cost to themselves and to prepare for a career, often in science or a technical field. More significant, perhaps, was the massive infusion of federal dollars into American higher education. The GI Bill covered tuition, or at least most of it, while the cold war forced Washington to invest considerable funds into solutions to military and other problems. In 1949, for example, the Office of Naval Research spent $29 million to fund more than 1,100 projects at more than 200 institutions. These moneys went to build laboratories, the scientific and technological infrastructure of universities and colleges; to hire more science faculty; and to fund additional research assistants.

Congress moved in 1950 to sustain this scientific and technical emphasis by creating a new federal agency, the National Science Foundation (NSF). Established to "develop and encourage the pursuit of a national policy for the promotion of basic research and education in the sciences," the foundation was headed by a director with staff but served under a president-nominated and Senate-confirmed board of science and technology experts, who served for six or possibly twelve years. Nominations for the various board members were sent to the president from scientific societies and professional organizations to ensure that the board "reflect[ed] the views of scientific leaders in all areas of the nation." Traditional national defense concerns found a prominent place in the new agency. The secretary of defense could command it to undertake or analyze defense-related scientific work but the foundation could not sponsor nuclear energy research without the express approval of the citizen-led Atomic Energy Commission. But while it could not sponsor nuclear research, the NSF could evaluate and coordinate scientific and technological efforts throughout government, and with private and other public groups. As a central clearinghouse of scientific and technical research, it gathered lists of American scientists and technologists and identified what research was being done in the various fields. The agency fostered scientific and technical interchange in America and elsewhere, but international cooperation required a statement by the secretary of state that the effort was "consistent with the foreign policy objectives of the United States." Awards were distributed solely on the merit of the proposal but recipients needed to sign a loyalty oath, guaranteeing in this bifurcated world that their allegiances lay with the United States.

Clearly, then, this new scientific agency was to be a vital piece of American foreign policy. It not only funded beginning and ongoing scientific and technical work, but the NSF also was responsible to create the next generation of scientists. By awarding "scholarships and graduate fellowships in the mathematical, biological, engineering, and other sciences," it could manipulate the number of persons engaged in scientific and technological inquiry, and, since the nation's military prowess depended on scientific and technical application, assure that America remained strong.[5]

Creation of the NSF was a departure from the status quo. It promised to open a new front on the knowledge frontier, to expand the means and direction of federally supported science work. But its initial minuscule budgets—no more than $5 million out of science-related federal expenditures of more than $2 billion per year—pointed in another direction. Its true significance was less practical or scientific than emotional. The NSF was a sop to those scientists who disliked military or applied research on moral or intellectual grounds or demanded scientist-controlled research agendas. In the first instance, it provided the possibility of research money untainted by crass militarism or materialism. In the second, its dependence on representatives of professional societies for selection of its administrative board assured that scientists would direct its endeavors. In some ways, the NSF was a perfect paper resolution to the World War II dilemma of how to configure postwar science in an age of atomic power and in an era when the nation's future depended on science. By reaffirming the principles held dear to scientists fearful of the military engulfing American science or disillusioned by their part in the creation of the atomic bomb, the NSF had a soothing influence. It assuaged guilt and gave the appearance of power sharing, sanctifying "original" or "pure" research and placing it in the hands of scientists.

GOVERNMENT ENTERS THE SCIENCE BUSINESS FULL-TIME

But this resolution was mostly smoke and mirrors. Far more significant was the fact that federal science research funding exploded after World War II and that the blast went in specific, certain directions. By far the greatest beneficiary was military or defense-related science. The Korean conflict gave the funding situation additional immediacy. In 1953, for instance, military/defense-focused science captured more than $1.5 billion of federal science's $2.2 billion budget. Central to these scientific endeavors were private contractors—corporations that bid for and signed agreements to undertake certain scientific work or to produce a particular military technology. But even more critical was the establishment of national laboratories. An offshoot of World War II, the national laboratories gained their impetus from the cooperative enterprise that constituted the Manhattan Project and other similar war-related, science-based activities. The federal government found the laboratories' wartime efforts so compelling and necessary in directing a science-based future that it converted these temporary entities into permanent establishments. But it did not run them itself. Instead, government leased operation to private contractors. For example, Oak Ridge National Laboratory, which had labored during the war to separate isotopes of uranium to gain enough fissionable material to make an atomic weapon, was leased to DuPont, then Monsanto, and finally Union Carbide. The Ames Laboratory, which had worked to separate uranium from other elements, was contracted to Iowa State University.

In conjunction with these new laboratories and others, a new type of entrepreneur emerged: the scientific administrator. Generally a senior scientist with a creditable research record—the Manhattan Project's Robert Oppenheimer provided an apt model—the administrator's job was to see that the money flow continued unabated, a crucial, but thankless position. He had responsibility to attract top-notch scientists, to keep them satisfied, and to keep scientific efforts on track, all the while riding herd on budgets. He needed to cut expenditures and increase future revenue to preserve the enterprise while at the same time fostering scientific initiatives. This managerial functioning required personal suasion and scientific aplomb as well as a keen and aggressive sense of where the possibilities lay and how to achieve them.

National laboratories and defense-related science gobbled up fully 68 percent of the federal science budget. The percentage of the moneys that went to the physical sciences was much, much higher. About 95 percent of all federal research dollars paid for physics, chemistry, or engineering. Only 5 percent went to the life sciences and medicine, and a considerable amount of this biological science money went to support agriculture. The remainder (!) went to the social sciences, which included funding for in-house efforts, such as at the Bureau of Agricultural Economics within the US Department of Agriculture (USDA), and contracting with outside "think tanks," such as the RAND Corporation, which was created to capitalize on Defense Department issues.

The new federal military and physics dollars proved particularly attractive to colleges and universities, even those without strong science research traditions. If science was the gateway to the future, then scientific research was a key to an educational institution's remaining relevant. And if funding was required to build strong laboratories—places in which to conduct research—then tapping into federal largesse was essential. And since laboratory work required many hands, entire new generations of scientists could apprentice in these new masters-research facilities. Certainly college and university scientists were familiar with the concept of federal sponsorship; virtually every American physical scientist had pursued science under federal auspices during World War II. As a consequence, schools long known for providing a strong liberal arts education now moved forcefully into federal-sponsored science research. Rarely did these initiatives begin with rank-and-file faculty. Almost always college presidents, boards of direction, or alumni dictated the transformation. They generally hired scientific administrators to spearhead the effort and to drag along reticent scientific faculty.

Princeton University was among those institutions pursuing the new manna. Like many other places, Princeton had been exceedingly active in World War II science. After the war, its administration aggressively added physical scientists, particularly aeronautical engineers, who had made significant reputations for themselves during the war. To support these experienced but new to Princeton researchers, as well as those who had returned to Princeton after the war, the university went after federal aeronautical contracts. By 1949, Princeton had federal research moneys of nearly $1.2 million compared to its total external research budget of only $30,000 some nine years earlier. While some at the university argued that the new emphasis on federally sponsored research had fundamentally changed the character of the place, the beneficiaries of federal funds argued that the "inflated costs of nearly everything" led them to embrace government funds "with considerable enthusiasm," a necessary facet of postwar science. In fact, the university's administration tried to reduce concern by arguing that the increase in federal dollars would actually humanize Princeton by enabling the university to use a greater percentage of its endowment for the humanities.

By the early 1950s, Princeton began to solidify its research efforts. Its administration recognized that since air power seemed a key to national security, aeronautical research was likely to bring in a virtually never-ending yet continuously growing federal revenue stream. Corporate donors helped the university build a new aeronautical research center—the James Forrestal Research Center—but a federal contract paid for its operation and research for fifteen years.

Through this period, Princeton had stepped up its research exponentially but sought to maintain a sense of balance by refusing to accept classified projects. This did not sit well with some physical scientists who threatened to leave the institution if the university did not lift its classified ban. President Harold Dodds soon weighed in on the side of the research scientists, casting the situation as a matter of academic freedom and arguing that classified research was "appropriate as a means of holding our departments together." By 1952 the university ended its classified prohibition "to prevent our faculty from being forced . . . to go elsewhere to work on important national problems or research in . . . their major interests." Federal grants then topped $3 million per year. The rest was clear. The institution had decided that its survival lay in federally supported physical science research. To continue to reject classified research money was to render Princeton irrelevant, its faculty and administration even unpatriotic.[6]

Princeton showed that scientists could disagree over the source of research moneys. But they and the vast majority of Americans in the early 1950s had little doubt that science remained the key to future security and prosperity. Science would overcome disease, lead to a higher standard of living, and ease the plight of humanity in the modern world. Nuclear power would produce energy too cheap to meter. Each garage would house a nuclear airplane and a nuclear car. And from this optimistic article of faith came a certain naiveté about the character of scientists. Since science would yield noble ends, its practitioners seemed above the fray, operating according to pristine motives. Such was the conviction of then President Eisenhower, who announced the Atoms for Peace program to the United Nations in 1953. The guts of the program were to make available to interested scientists throughout the world fissionable material to "encourage worldwide investigation into [nuclear power's] most effective peacetime" usage. Eisenhower showed little concern about security among scientists; he did not worry that some scientist might take this weapons-grade material and use it for nefarious purposes. He did fear that criminals or rogue states might capture the material, yet he expressed his confidence that science would also solve that problem. "The ingenuity of our scientists," Eisenhower noted, "will provide special safe conditions under which such a bank of fissionable material can be made essentially immune to surprise seizure."[7]

America, then, had moved headlong to bolster its science program. Dominated by funding for problem-directed research, the nation made some small allowances for scientist-determined research questions and seemed right on track. Federal expenditures on science increased an average of 15 percent yearly even as the economy was suffering a serious slowdown. In an effort to make the direct-basic research distribution even more appealing to the nation's scientists, the federal government required its established agencies to engage in what became known as "mission-oriented basic research" even as they pursued specific questions. In 1956, for example, agencies such as Defense, Agriculture, and the like were authorized to spend nearly $200 million on this oxymoronic concept. Defense "mission-oriented basic research" comprised about 60 percent of the total. In his charge to laboratory directors, the USDA's Byron T. Shaw defined mission-oriented research and showed its practical bent and dramatic difference from "pure" science. Mission-oriented research, Shaw said, is "undertaken to discover the principles underlying research areas and to develop theory which will greatly facilitate problem research as needs arise." Such research, Shaw concluded, "will be expected to build a foundation for the quick, effective, and economic solution of research problems."[8]

THE FUTURE IS LESS BRIGHT: SPUTNIK AND AFTER

The sense of success and optimism was shattered on October 4, 1957, when the Soviet Union launched *Sputnik*, the world's first artificial satellite. This extraordinary scientific and technological achievement startled Americans, who had furiously been working on their own rocket program. Indeed, *Sputnik* seemed so dramatic and potentially menacing, not because America had not mobilized its science and technology efforts, but rather because it had. Despite institution-alizing the lesson of World War II that science and technology were the keys to the future and that the federal government needed to pursue them vigorously, democracy's mortal enemy—communist totalitarianism—beat the United States into space. This suggested something much greater than a lead in rocketry. It implied that communism might be superior to capitalism and that the productive capacity of communism would soon overwhelm the United States.

Not surprisingly, communist regimes celebrated the event. East Germany proclaimed that "the technical lead of the Soviet Union . . . prov[es] the superiority of socialist society and production over that of the capitalist world." The People's Republic of China declared that "the United States has hitherto bragged that it was the most powerful on earth. But now it is lagging behind the Soviet Union." Nikita Khrushchev, premier of the Soviet Union, while noting that "we have passed the richest and most advanced country of the capitalist world—the United States," signaled that the new satellite also gave the Soviet Union a tremendous military advantage. "They might as well put bombers and fighters in the museum," he concluded.

Voices in the United States tried to downplay the Soviet scientific achievement and to relieve the fears of the populace. The *New York Times* argued that American scientists "rank often at least equal" to their Soviet counterparts and that the Soviet satellite program depended on "heavy taxation," which "results in a lower standard of living for the average Soviet citizen." President Eisenhower contended that the United States could have launched a satellite first but unlike the Soviets refused to merge its scientific and military enterprises. To effect such a dramatic union would have led "to the detriment of scientific goals and military progress."

Eisenhower conceded that the Soviets did gain "a great psychological advantage in world politics," a view with which the *Times* agreed. It called for more funding, and the Pentagon announced that it was assigning our top scientists to the satellite project. Eisenhower gave immediate approval for budget overruns in satellite and rocket programs. He also created the President's Science Advisory Committee to

devise solutions to the defense-related science problems that characterized the post–World War II world. A Senate probe into what was
wrong in the American program began two days after *Sputnik*'s launch.
Vice President Richard M. Nixon was more straightforward. *Sputnik*
showed, he urged, that unless the United States provided considerably
more funding and more manpower to scientific and technological
questions, the USSR would surely overtake America.

In this heated atmosphere, the question was not whether but
when and how science funding would expand. Predictably, scientists
engaging in the public discussion argued that to focus only on one
sector—satellites—was to do the future a profound disservice. Only
a broadly gauged effort to create a nation of scientists could hold off
the communist menace. "We must generate the will to supremacy,"
argued Wernher von Braun, the United States' leading rocket scientist. "The modern world," contended von Braun in testimony before
Congress, requires "a new kind of soldier . . . the man with the slide
rule. The 'national interest' demands 'that we increase the output of
scientific and technical personnel.' It is the role of 'the federal government' to create this cadre not by 'dictat[ing] such a program' but
by assuming financially great 'pump-priming role in the public
schools and in our colleges and universities.'" James R. Killian, special assistant to the president for science and technology, chair of the
President's Science Advisory Committee, and president of Massachusetts Institute of Technology, took an even more aggressive
stance. He urged Congress to increase dramatically federal science
funding, to foster a greater emphasis on the physical sciences in
public schools, and to change the manner in which children were
taught science in secondary schools. The last of these could be done
"by building into [science teaching] some of the intellectual rigor
and excitement, some of the beauty and humanistic values inherent
in modern science." Claiming "a man cannot be really educated in
a relevant way for modern living unless he has an understanding of
science," Killian called for a national campaign to bring "scientific
literacy among the rank and file." (See fig. 2.)

C. V. Newsom, president of New York University, a university
lacking a significant science or technology tradition, disagreed

http://history.nasa.gov/sputnik/rocketholding.jpg

Fig. 2

Amid cold war political and military tensions, the Soviet Union's launch of *Sputnik* in October 1957 seemed to embarrass the political, military, and engineering establishment of the United States. America's first attempt to salvage its dignity by quickly launching its own satellite in December 1957 ended abruptly in technical failure. The United States finally managed to send up its own satellite, *Explorer I*, on January 31, 1958, aboard a different type of rocket. *Explorer I* made a statement; while critics derided *Sputnik* as showy but useless, simply a little ball going beep-beep, *Explorer I* carried scientific instruments designed and built by James Van Allen of the University of Iowa (center figure). Using data collected by *Explorer I* and subsequent missions, Van Allen proved that powerful belts of cosmic radiation encircled the earth. Van Allen is shown here with JPL lab director William Pickering (left) and rocket engineer Wernher von Braun (right) displaying a model of *Explorer I* after its successful launch.

somewhat. He contended that the key to future success depended on creating a new generation of "men and women who have powers of imagination and creativity," people who will "promulgate new ideas of great significance." Education must aim beyond just science to allow us to "make progress in the resolution of our social, political, moral, and economic problems and in the continued improvement of our standard of living."[9]

Lobbying was furious within the White House and Congress. In the end, Congress fashioned a law that spoke to the broad and often contradictory interests of the administration, scientists, educators, Democrats, Republicans, and other groups vitally involved in responding to *Sputnik*. The National Defense Education Act (1958), or NDEA, was the result. The law began with the grave declaration that "an educational emergency exists and requires action by the federal government" and that Washington must and will "help develop as rapidly as possible those skills essential to the national defense." Disavowing any interest in dictating or controlling curricula, the act provided federal moneys for every level of the American educational system. Perhaps its most significant mandate established a threefold approach to colleges and universities. First, it provided in a four-year period guaranteed loans of roughly $300 million at nominal—3 percent—interest to college and university students to increase greatly the numbers of students who could attend college. Preference for loans was given to academically superior students who chose to teach in elementary or secondary school or who majored in "science, mathematics, engineering, or a modern foreign language." The act budgeted a similar amount for enhancing science, math, and foreign language instruction at these institutions as a way to expand facilities for scientific work. Finally, 5,500 outright three-year fellowships were granted to students attending new or expanded graduate programs with preference given to those intending to become college teachers as a means to prepare the next generation of researchers and professorate.

These initiatives provided a huge infusion of federal funds into higher education. The comprehensive law did not neglect secondary and primary education. It also authorized significant expenditures

to create and train a cadre of guidance counselors for the public secondary schools of each state, to identify able students and encourage
them to attend college. It funded efforts to find ways to utilize more
effectively in education the newer technologies of television, radio,
and motion pictures, hoping to stimulate creativity and intellect. It
established vocational education programs in secondary schools to
train technicians in science and technology "to meet the needs of
national defense." And finally, the act provided for the establishment of institutes where people could learn the newest methods of
teaching the modern languages—French, German, Spanish, and
Russian, languages necessary to wage and win the cold war.[10]

The NDEA then addressed thoroughly the question of how to
ensure the regular and persistent production of scientists and engineers. A second initiative undertaken in *Sputnik*'s wake had a more
nebulous charge but an equally important mission. Early in 1958 a
Department of Defense directive created the Advanced Research Projects Agency (ARPA). This new body was formed "for the direction or
performance of such advanced projects in the field of research and
development as the Secretary of Defense shall, from time to time,
designate by individual project or by category." Under this amor-
phous, secretive directive, the ARPA was to think, plan, and execute
bold new scientific and technological enterprises. The shadowy ARPA
undertakings were to be grand strikes, explicitly not suited to immediate defense needs. In fact, ARPA did not even target the next
generation of science and technology but sought to skip several generations of thought, experiment, and product. Its parent agency
understood that many of ARPA's ideas would not pan out. But the
possibility of a quantum leap in science and technology, a jump similar perhaps to that yielding atomic weaponry, promised to redefine
the basis of the scientific and technological contest between communism and democracy. Simply proving the technological feasibility of
a process and developing a strategy for creating a prototype system
could result in a breakthrough of unprecedented importance.[11]

ARPA's initial project was to oversee and direct the future of the
US military missile effort. But like the NDEA, the ARPA was not to
serve immediate needs. A third new program came to bear directly

on *Sputnik*. To speed America's satellite efforts and to maintain the fiction that the space quest was primarily an intellectual, not military, enterprise, Congress folded the National Advisory Committee for Aeronautics into a new agency, the National Aeronautics and Space Administration (NASA). Much more generously funded than its predecessor and remaining nominally under civilian control, the new agency aimed to signal to the world that America, unlike the USSR, championed the peaceful exploration of outer space. In NASA, scientists led the charge to space, not military men, and so demonstrated the difference between the individual expression of American freedom and the collective, centralized, bureaucratized rigidity of communism. In that spirit, the agency supported efforts to learn about the upper reaches of the earth's atmosphere and to plan satellites to help predict the weather. But it also ventured into areas more commonly associated with the military. NASA developed its own thrusters and sought to create nose cones easily retrofitted for warheads.

NASA mimicked a federal research approach that had found its greatest expression in the atomic research of World War II, especially the quest to achieve a concentration of fissionable material dense enough to sustain a chain reaction. There, as in NASA, the important point was planned duplication and competition. Several agencies, groups, or teams undertook the same or similar tasks. Yet they were left primarily to their own devices to achieve those tasks. This competition, to some, seemed wasteful, not necessary. But others argued that several sets of researchers working on similar questions actually saved money by speeding task completion and by all but guaranteeing that the best solution would surface.

In the year or so after *Sputnik*, the United States had redrafted and redoubled its scientific enterprise. Through federal funds, it established means and methods through which America would better wage the cold war. The results were dramatic. The US government's science budget doubled between 1957 and 1961 to more than $9 billion. In 1960 the United States spent a greater proportion of its gross national product on science than any other nation. The vast majority of the moneys went directly to resolve specific problems, almost all of which revolved around defense. Also in 1960 the basic research budget, including mission-oriented basic research, topped out at 7 percent. While the NSF budget expanded more than tenfold from its initial 1950 appropriation, its percentage of the relatively meager basic research budget never eclipsed 15 percent.

Imitation became the sincerest form of flattery as every move by the Soviets seemed to demand an immediate response by the Americans. When he became aware that the Soviet Union was planning a huge new synchrotron, Eisenhower announced his support for an even larger machine to be built in America and asked Congress for $100 million to start the project.

Eisenhower's proposal was more symbolic than substantive. Like placing NASA under civilian control, Eisenhower aimed to remind the world that the cold war was truly a broad-based contest between ideologies. Those ideologies were embedded in and infused throughout the character, institutions, and expectations of the respective societies. To Americans in the 1950s, choice, plenty, and flexibility were democracy's outward expressions; they translated into the freedom to choose and the opportunity to have choices to choose from. Rather than having the state dictate an activity or product, Americans demanded that the federal government help preserve the freedom and ability of choice. In effect, the American Dream of the 1930s became choice in the 1950s; the American way of living no longer meant modernity but the option of choosing.

The federal government's methods for ensuring, even expanding freedom to have choices to choose from, were fostered through science and scientific applications that could be adapted by large or small corporations. The government's methods were a natural outgrowth of American expectations for scientific and technological choices—options that were as much a part of federal science and technology policy as defense was against the communist menace. In agriculture, for example, the USDA created an Agricultural Research Service (ARS) in 1953 explicitly to consolidate and focus research efforts. In the past, USDA research had concentrated on some immediate blight or animal disease or new means to harvest a crop. But from this new group came broad innovation, including thousands of new strains of plants and animals; studies in the chemistry of aromas and flavors; use of anhydrous ammonia, pesticides, and herbicides to increase crop yields without submitting fields to periods of fallow; addition of antibiotics to the feed of production animals; studies of the structure of estrogenic substances in various clovers; creation of instant mashed potato flakes; and development of fire retardants for garments.

This research thrust paralleled the rise of processed foods in the years after 1950. Corporations worked to produce modern foods: they sought emulsifiers, flavor enhancers, and spoilage retardants to free American families from the tyranny of fresh foods. Private enterprise labored to reduce costs and to create products that would encourage flexibility in American life by freeing home cooks from the natural, restricting parameters of foods in their natural state. This was the way to increased profits for producers and satisfaction for consumers. In the 1950s manufactured foods had increased shelf life, improved taste, and stabilized quality. Manufacturers' aim was to create a consistent, dependable, desirable product, not subject to the whims or capriciousness of nature. Fish sticks, constructed from minced fish, appeared first in 1952, along with potpies; dehydrated onion soup; powdered, nondairy coffee creamer; and artificially sweetened soft drinks. Frozen dinners—TV dinners—were launched a year later. Instant iced tea mix was offered in 1953 and Cheez Whiz became a substitute for cheese in the same year. Self-basting turkeys were commercially available in 1954. In addition to convenience foods, fast food—food modified and manipulated to provide consistent taste, stability, and longer shelf life, and made outside the home so as to require little or no familial preparation and cleanup—became commonplace. Dunkin' Donuts, Denny's, Burger King, Kentucky Fried Chicken, McDonald's, International House of Pancakes, Pizza Hut, and Jack in the Box all first tempted the American palette in the 1950s.

Consumers profited in another way from the research thrust. In 1956 Congress called on ARS to greatly expand research into new farm product use. The reasoning was simple: farm crops and their stalks, roots, leaves, and the like were huge sources of chemicals. Chemicals derived from farm products might enable manufacturers to replace imported ones and therefore increase availability of some crucial products within the nation. Farmers, meanwhile, would get new markets for their crops. Using grain as motor fuel was among the most successful ventures, but the USDA also developed processes to render plants and animals into the manufacture of plastics, adhesives, and even pharmaceuticals. Congress also demanded that the USDA increase its grant program for students majoring in agricultural science, expand government-industry cost sharing in the application of ARS science, and offer economic incentives to get industries to initiate large-scale pilot programs based on ARS science.

The ARS's efforts were also felt outside the United States. After the passage of the Agricultural Trade Development and Assistance Act, the Foreign Agricultural Service (FAS), a new agent of the USDA, carried the fruits of federal science to America's friends and unaligned nations. On the most basic level, FAS took advantage of the federal science–fostered agricultural surplus—while the Soviet Union had difficulty feeding its people—to demonstrate the superiority of the American agricultural system, and made the surplus available as foreign aid to nations facing food emergencies. Less desperate nations lacking food sufficient for their populations could barter material desired by the American government in exchange for surplus grain. In 1959 these initiatives had all been combined under the slogan "Food for Peace." The FAS also established a program for foreign students to come and gain training in American agricultural science. Aspects of this training included how their knowledge might be implemented to overcome the problems of their native countries. Through international agricultural trade fairs, another facet of the FAS, American science and technology stood side by side with its Soviet counterparts, permitting those in attendance to compare the sleek, modern American devices and products with their more cumbersome, bulky Soviet complements.

Though not involved directly with agriculture, a political version of this comparison was aired on American and Soviet television during the famous 1959 Kitchen Debate. In the USA exhibit at a USSR industrial exhibition, Khrushchev and Nixon engaged in a heated argument over the virtues of their nations' consumer products and the ideologies manifested in them. Nixon showed the premier a modern American kitchen, a color television, a data-processing machine, an automatic floor sweeper, and other consumer products. At each location, the two sparred over which system provided their citizens with the best prospects and how military prowess and industrial production related to scientific accomplishment. When they reached a built-in, panel-controlled washing machine, Khrushchev could no longer contain himself. "Don't you have a machine that puts food into the mouth and pushes it down?" he barked. The things Nixon had shown "are not needed in life . . . They are merely gadgets." Nixon dismissed Khrushchev's comment and reminded the premier that in the United States "we have many different manufacturers and many kinds of washing machines so that the housewives have a choice." Our aim at the exposition, he continued, was not "to astonish the Russian people. We hope to show our diversity and our right to choose." That was the fundamental difference between democracy and communism, Nixon concluded. "We do not wish to have our decisions made at the top by government officials who say that all homes should be built in the same way."[12] (See fig. 3.)

Nixon's defense of the American system did not sway Khrushchev, of course, and it also failed to convince a small segment of elite American society. These men and women worried that the United States was infatuated with material goods and the creation thereof, and scolded other Americans for selling their souls and spirits for refrigerators and other property. They felt that traditional endeavors —intellectual, artistic, and cultural—were being shunted aside in a mad dash for the newest, most novel device. These men and women used the term *gadget* derisively, to signify anything they deemed trivial. This critique included an attack on the concept of consumer. While some maintained that individual ownership and security "may be the highest reach of civilized life . . . within the grasp of common humanity," detractors claimed that financial wealth was being confused with spiritual well-being.[13]

Private enterprise used federal science to produce consumer commodities. This easy, hand-and-glove relationship permitted manufacturers to translate into products techniques developed with federal

http://history.sandiego.edu/cdr2/USPics/82463.jpg

Fig. 3

In July 1959 American vice president Richard Nixon visited Moscow and engaged Soviet premier Nikita Khrushchev in what became known as the "Kitchen Debate." While visiting the American National Exhibition held in Moscow, the two leaders disputed the strengths of capitalism versus communism by comparing how the two systems provided ordinary citizens with material comforts and labor-saving devices. The exchange made good political theater but also revealed the underlying tensions over the cold war military arms race.

funds, often in government-financed agencies with research done by federally employed scientists. In this scheme, the federal government was simply fulfilling its democratic responsibility, to promote choice, options, opportunities. Provision of these alternatives appeared as a bulwark against communism, making America a beacon of freedom and plenty to the world. Yet with this new scientific responsibility came a new federal scientific obligation. The federal government must regulate the application of science to commercial goods.

The US government long had regulated private enterprise in the public interest. The Constitution granted the federal government power over interstate trade. Antimonopoly statutes and campaigns dominated turn-of-the-century politics. Adulterations (products that were not what they claimed to be or that produced harm) and truth in advertising (making claims for a product that matched its capabilities) had interested New Deal–era Americans. The thrust of the 1950s was somewhat different. Emphasis on the democracy of choice posed a conundrum: On what basis does one choose? Modernity? Convenience? Certainly, convenience was a relative, individualistic term. Some might opt for one type of product with specific parameters and others might gravitate to different configurations. Beyond convenience, however, the case was more difficult. The fact that many new products and substances were science or technology based exacerbated the matter: products may not have been what they seemed.

Many persons complained of their inability to choose wisely among the many new products available to them; they lacked the knowledge or technique necessary to safeguard themselves and their loved ones. From this dissatisfaction and confusion coalesced a new interest group: consumers. Generally moderately well-to-do female homemakers, these people worried less about value than salubrity and demanded that government protect their families' interests.

Nowhere did protection for families seem more urgent than from the food processing industry. Government and other scientists had produced the knowledge necessary to create chemical emulsifiers, stabilizers, and flavor enhancers, which freed consumers from fresh foods' limitations. Increasingly in the 1950s and especially after 1956, consumers and their supporters called for the federal government to regulate food additives, to determine if any chemicals posed a threat to an individual's health. Consumers were delighted by choice but recognized the federal government's obligation to control the healthfulness of manufacturers' offerings.

Congress complied in 1958 by passing the Food Additives Amendment. This epic act institutionalized scientific regulation within the federal bureaucracy. At its heart, the act rendered regulation a matter of probabilities, hypotheses to be tested, works in progress. In that sense, it mimicked science, which tested ideas and theories, always remaining open to new data, results, or arguments that might prove an established process or procedure erroneous.

The amendment established the Food and Drug Administration (FDA) as the agency in charge of food additive regulation and designed two basic procedures. Defining a food additive as anything used in the producing or processing of food, the amendment employed a sort of scientific plebiscite to consider food additives in use prior to January 1, 1958. "Experts qualified by scientific training and experience" were polled "to evaluate its safety" for each additive "in common food use" prior to that date. Every additive that was judged "safe under the conditions of its intended use" and that therefore received a relatively clean bill of health and an endorsement from the scientists would continue in the food supply as a Generally Recognized As Safe (GRAS) substance. GRAS material could later be reclassified only if "scientific procedures or experience" demonstrated that the material constituted a public health hazard.

Results of the scientific plebiscite were tabulated and evaluated by FDA scientists, who became the linchpins of the new regulatory effort. They also ruled on the legality of new food additives. FDA scientists understood that food manufacturers could not guarantee safety or harmlessness, of course, so they required their corporate counterparts to do the next best thing: before they could introduce their new creations into the food supply, they needed to demonstrate scientifically to FDA scientists that within the bounds of then-contemporary scientific knowledge the additives in question were likely not harmful. Again, if new scientific information surfaced, FDA scientists could revisit their approval.

A potent assumption lay at the base of this new regulation. Food additives were the products of scientists in industry; and as scientists, these men and women talked the same language as those in government, shared the same values, and had the same reverence for scientific truth and the scientific method. In this view, regulation was not an adversarial relationship but rather a partnership in which scientists both within industry and within government worked together to reduce hazards to the public. Without that understanding, regulation as it was fashioned in the 1950s could not be successful. FDA scientists had neither the equipment and facilities nor the staff to perform and verify the vast majority of industry scientists' contentions. But such a procedure seemed unnecessary. As in the claims undergirding the Atoms for Peace initiative, it seemed as if scientists—because they were scientists and trained in a specific rigorous methodology—viewed data and truth similarly and worked together free of outside forces.

In only one instance did the Food Additives Amendment take the matter out of scientists' hands. It included a per se clause with respect to cancer. No additive "was deemed safe if found to induce cancer when ingested by man or animal." Virtually no scientists favored the measure, including those responsible for regulatory matters or those affiliated with the National Cancer Institute. They disliked substituting a per se clause for scientific analysis—the kind of analysis institutionalized in GRAS—and they thought it unenforceable because they objected to the idea of proving cause. To prove something scientifically as the cause of something else was a most difficult challenge. It required more than a statistical correlation, forcing investigators to rule out virtually every other possibility before cause could be established.[14]

Such a clause did little to protect the public, they argued, and it only made their labors more difficult. Their arguments proved prophetic when in 1959, just two weeks before Thanksgiving, a small portion of the Oregon cranberry crop was found to be tainted with a minute but identifiable quantity of aminotriazole, which was associated with an increase in thyroid cancer in rats. Arthur Flemming, head of Health, Education, and Welfare (HEW), the FDA's parent agency, announced the discovery; and although his scientists found no other incidence of contamination, Flemming warned the nation's homemakers that to be on the safe side, they should forego cranberry sauce at this year's holiday dinner table.

Cranberry producers outside of Oregon were incensed at Flemming. So too were most scientists, who noted that the contamination in Oregon was infinitesimal and that to achieve the dosage that was used to induce cancer in rats required a human to ingest fifteen thousand pounds of berries a day for many years. Supermarkets quickly removed cranberry products from the shelves, which threatened to punish all cranberry growers. Several states and cities banned cranberry sales.

The *New York Times* led the counterattack, capturing the scientists' contention. It was argued that even if the public consumed only Oregon cranberries—they remained banned throughout this episode—the quantities necessary to do harm were stupendous. Political leaders joined in. Ezra Taft Benson, the secretary of agriculture, announced that despite Flemming's warnings, no menace existed and that he planned to eat cranberries at Thanksgiving. Vice President Nixon, beginning his own presidential campaign, ate four helpings of cranberry sauce on campaign stops to demonstrate his belief in the lack of danger. Not to be outdone, his soon-to-be Democratic rival, Massachusetts Senator John F. Kennedy, helped himself to two glasses of cranberry juice on the campaign trail. The pressure on Flemming soon became overwhelming and he announced conspicuously that he too would eat cranberries at his traditional Thanksgiving feast. Soon thereafter, cranberry products reappeared in supermarkets just in time for the holiday.

To scientists (and cranberry growers, albeit for different reasons), the damage had been done. A per se law had agitated the public and needlessly usurped scientific prerogative and mediation. It placed scientists in the position of seeming to support introduction of a carcinogenic agent into the food supply and demeaned them in the public estimation. To many scientists, it appeared a gratuitous rejection of science and scientists by government, especially Congress, and they believed that institutionalization of the per se clause did nothing more than reflect political partisanship at the expense of the public weal.

ELECTION OF 1960: EMERGENCE OF KENNEDY

With Eisenhower ineligible to seek a third term, the 1960 presidential election pitted Nixon against Kennedy. World communism, the atomic bomb, and *Sputnik's* implications helped define the contest. So too did the U-2 incident. An American spy plane was shot down over the Soviet Union on May 1, 1960. At first Eisenhower denied the event, but the Soviets had captured the pilot and put him on display. Eisenhower quickly admitted the deception and relations between the two superpowers chilled precipitously. That the Soviet military was capable of shooting down a plane designed to fly at seventy thousand feet startled Americans. The nation seemed even more endangered by Soviet military prowess than it had thought.

Nixon capitalized on this theme. "We are in a race tonight, my fellow Americans, in a race for survival in which our lives, our fortunes, our liberties are at stake," he barked. He recognized that "we are ahead now," but maintained that "the only way to stay ahead in a race is to move ahead." Certainly the Eisenhower administration had made great strides to keep the nation safe. It was wrong to charge "that American education and American scientists are inferior." It was wrong to assert "that America militarily and economically is a second-rate country." But more needed to be done. Americans in all walks of life needed to redouble efforts. "Students in schools" needed to strive "for excellence rather than mediocrity." "Every boy and girl of ability regardless of . . . financial circumstances" must "have the opportunity to develop his [sic] intellectual capabilities." Scientists must receive "the support they need for the research that will open exciting new highways into the future." "In this decade of decision and progress," Nixon contended, we need to "witness the continued revitalization of America's moral and spiritual strength with renewed faith in the eternal ideals of freedom and justice." Yet Nixon made it perfectly clear that he understood traditional Republican precepts. Nixon believed that the government has a tremendously important role in affecting the future blossoming of America, "but the role of government is not to take responsibility from people but to put responsibility on them." Government should "not dictate to people but to encourage and stimulate the creative productivity of . . . free Americans." We are in desperate straits, he concluded. "The question now is freedom for all mankind and the survival of civilization."[15]

Nixon pointed with pride to the Eisenhower administration's accomplishments but recognized a potent need to extend them. Kennedy took a much different tack. He rejected any thought of continuing the Eisenhower approach, stating that Nixon's program "is a pledge to the status quo—and today there can be no status quo." "This nation cannot afford such a luxury," he claimed, as "the times are too grave, the challenge too urgent, and the stakes too high." Already "the balance of power is shifting." "Friends have slipped into neutrality—and neutrals into hostility." During the Eisenhower administration, "there has been a change—a slippage—in our intellectual and moral strength. Seven lean years of drought and famine have withered the field of ideas . . . A dry rot, beginning in Washington, is seeping into every corner of America."

Rather than "look to the safe mediocrity of the past," he urged Americans to embrace what he called "the New Frontier," to welcome "a new generation of leadership—new men to cope with new problems and new opportunities." The "times demand invention, innovation, imagination, decision." Can "our society—with its freedom of choice, its breadth of opportunity, its range of alternatives . . . compete with the single-minded advance of the Communist system?" Can the United States "carry through in an age where we will witness not only new breakthroughs in weapons of destruction—but also a race for mastery of the sky and the rain, the ocean, and the tides, the far side of space and the inside of men's minds?" The election, Kennedy deduced, was quite simple. The nation must choose "between public interest and private comfort—between natural greatness and national decline—between the fresh air of progress and the stale, dank atmosphere of 'normalcy'—between determined dedication and creeping mediocrity. All mankind waits upon our decision. A whole world looks to see what we will do."[16]

Kennedy's predominant objection to Eisenhower's leadership revolved around federal power. Eisenhower used government to promote freedoms and Nixon vowed to follow the same course. Kennedy argued that the new variables of the world made that approach obsolete and demanded now that government create basic human freedoms. In Kennedy's framework, science seemed destined to play a much larger part in federal initiatives. Kennedy's narrow election win—one of his most successful campaign slogans was that there existed a missile gap between America and the USSR, which he promised to close—seemed to guarantee that such a transformation would occur.

But before he left office, Eisenhower took a final opportunity to comment on the relationships among defense industries, universities, and the federal government. In this instance, he looked inward. Perhaps because he had been supreme allied commander during World War II and later president of Columbia University, a large research university, Eisenhower expressed considerable trepidation about connections between them. Moreover, he recognized that several of his presidential policies directly contributed to his worries. To fight the cold war, Eisenhower had helped "a permanent armament industry of vast proportions," different than anything he had encountered even in World War II, to take a central part in American life. Military security, he argued, now cost more than "the net income of all United States corporations." This new leviathan has a "total influence—economic, political, even spiritual—[that] is felt in every city, every state house, every office of the federal government." This defense-federal nexus constituted a fearsome "military-industrial complex," whose "potential for the disastrous rise of misplaced power exists and will persist." "Never," Eisenhower pleaded with his countrymen and women, "let the weight of this combination endanger our liberties or democratic processes."[17]

Eisenhower knew of what he spoke. Numerous companies depended heavily on government military science and technology contracts. Boeing, McDonnell, and Lockheed supplied the American military with thousands of bombers, jets, and transports. Boeing and Hughes also designed and built helicopters and missiles to launch satellites and intercontinental ballistic missiles (ICBMs). McDonnell made space capsules that would send early astronauts into space. Lockheed was the contractor for the U-2 and other spy planes. Raytheon specialized in manufacturing nuclear-tipped missiles to defend against an atomic attack. When incoming warheads were spotted, Raytheon's missiles were to intercept them in outer space or the upper atmosphere. Raytheon's rockets would detonate in the vicinity of these enemy weapons and blow them to smithereens, far from population or military centers. Nuclear weapons facilities were contracted to DuPont, National Lead Company, Lockheed Martin, Bechtel, General Electric, Monsanto, Union

Carbide, Goodyear, Westinghouse, Dow, and Mallinckrodt, among others.

Certainly there was a military-industrial complex entrenched in an incestuous marriage with the federal government. But Eisenhower also decried several other situations. As president, he had presided over part of "the technological revolution of recent decades." Eisenhower referred to atomic weapons and jet planes, of course, but he also acknowledged nonmilitary discoveries—synthetic foodstuffs, antibiotics, Salk and Sabin polio vaccines. Each of these and other facets of the technological revolution changed America, but they also changed how innovation occurred. Now "formalized, complex, and costly" research was the sine qua non. "A steadily increasing share"—the vast majority by 1960—of this research, said Eisenhower, "is conducted for, by, or at the direction of the federal government." In this world, the structure of the Manhattan Project had become the normative model. "Task forces of scientists, in laboratories and testing fields" have displaced individual scientists.

This was particularly destructive to "the free university, historically the fountainhead of free ideas and scientific discovery." Now, because of the huge costs involved in modern research, "a Government contract becomes virtually a substitute for intellectual curiosity." Such a condition "is gravely to be regarded." From now on, "the prospect of domination of the nation's scholars by Federal employment, project allocations, and the power of money is ever present," Eisenhower fretted.

Federal largesse may have tainted the university but reverence for science posed yet another threat. Since Americans so heavily held "scientific research and discovery in respect," citizens from all walks of life needed to prevent "the equal and opposite danger that public policy could itself become the captive of a scientific-technological elite." Rather than popular sentiment and electoral debate, pressure might be placed on people to forego their own judgment and to defer to a relative handful of technocrats, who might suspend democratic freedoms in their quest for scientific precision.[18]

No doubt Eisenhower thought of the Soviet system when warning against technocratic tyranny. In any case, he counseled caution for his country, pleading with its citizens to examine the institutional and intellectual framework his policies had helped create. When Kennedy took office, he was not nearly as circumspect. Science in the Kennedy administration assumed an even more exalted priority, and those scientists willing to apply their insights and techniques to public policy were appointed in unprecedented numbers.

Kennedy hoped to use the methods of science to fashion solutions to America's social problems, many of which stemmed from the application of federal science. He noted a "technological revolution on the farm," the runoff from which could taint the water supply, a concomitant "urban population revolution," a "medical revolution," which among other things increased life spans as well as incidences of diseases associated with aging, and a "revolution in automation," the last of which caused persistent unemployment. Kennedy refused to wait for private enterprise to attack these problems through free-market activities. For Kennedy, an affluent society had a sacred responsibility to undertake public interest projects, especially those that did not seem immediately profitable. The president demanded quick results, and groups of scientists in fields other than physics and chemistry descended on Washington to produce them. These men and women had made their reputations in academia and now laid claim to being among America's best and brightest.[19]

Kennedy emphasized biological sciences and actively engaged social science. His application of federally supported science to approach America's myriad social issues was a distinct reconceptualization of the nature of government science. "Is there no limit to the sweep of science and its useful application?" asked Jerome B. Wiesner, Kennedy's science advisor and a longtime science administrator on leave from the Massachusetts Institute of Technology's department of electrical engineering. "There may be," he concluded, "but its boundaries are certainly not visible from where we stand." Wiesner, Kennedy, and some congressmen discussed the possibility of creating a cabinet-level department of science during the first year of the new administration and again later but ultimately decided against pursuing it. While they contended that it would increase synergy among the many federal science initiatives, they worried that it might constrain scientists, forcing them to hew to a single model of activity and a single funding source. Such a draconian organization certainly would frustrate scientists, who enjoyed freedoms within the government similar to those of a free market; they could now approach the funding agency most sympathetic to their work. That entrepreneurial spirit, Wiesner and Kennedy believed, would be constrained within a single science department.[20]

Nor did established agencies and departments want to cede the right to fund the projects they chose and in the matter that they selected. To place these separate bureaucracies under a single science department would reduce their discretion and maneuverability. As significant, they could point yet again to the Manhattan Project model in which several groups of investigators competed to produce the first atomic weapon. Installing competition at the heart of the federal science effort might institutionalize redundancy, they conceded, but it also promised to speed work and produce the best possible result.

In lieu of a single science department, Kennedy, Wiesner, and Congress opted for an Office of Science and Technology. Located in the White House and staffed entirely with scientists, the office identified scientific issues and phrased questions for the president. Its charge included assisting the president on all matters of national policy impacted by or involving science and technology, coordinating the various federal science and technology functions—a representative worked within each agency or department to harmonize competing interests, working with the Bureau of the Budget and the president's Council of Economic Advisers, and serving as science liaison between the executive branch and Congress. Perhaps more than others, Wiesner recognized that the new social problem–oriented science might well face challenges quite different than defense-related science. Science in the service of national security had had almost universal acceptance. Science in the service of society ultimately pitted one group against another. It also raised the question of value. Would applied efforts be prized more than "pure" research? How might the tangible investments be measured? What about the nature of universities?

These questions concerned Kennedy, who became the first president since Lincoln to address the prestigious National Academy of

Sciences, but he thought them resolvable. Kennedy had great faith in scientists. He personally enjoyed talking with them and made sure that some graced every social gathering he held. But Kennedy had a very different outlook about the role of scientists within his administration than had Eisenhower. He did not wish to appoint impartial advisors; Kennedy selected only those committed to the success of his presidency and vision. That decision helped advice and its implementation to go hand and hand, but it also meant that the range of recommendations that Kennedy might hear would be tightly circumscribed. These parameters also reduced the number of scientists who might serve in his administration. Only policy intellectuals—those anxious to test their theories of human behavior by using the power of national government to mitigate social ills—were eligible, and then only if their prescription agreed with that of the president.

Certainly Kennedy made government service attractive. Among his first acts was to convince Congress to authorize 480 new "supergrade" science positions and an additional 280 regular ones to work in the various executive departments. Supergrade scientists received the highest pay in government. Perhaps his creation of the Council of Economic Advisers set the tone for New Frontier science. Each member had had a solid academic career but lusted for more; not to apply one's work to society struck them as sterile, reducing economics to irrelevance. All claimed that the national economy could be planned and managed. All had witnessed the managed economy of the 1940s and hoped to duplicate that success in the 1960s. Each supported the theories of John Maynard Keynes, believing that demand, not supply, was the key macroeconomic variable. Demand could be modified and encouraged, they argued, and the federal government could be an effective and proper agent of stimulus.

These New Frontier economists first made their mark when they attacked the sluggish economy in late 1961. Projections indicated that tax revenues would not meet anticipated expenditures and a deficit would result. Traditional economists counseled that the federal budget be balanced in peacetime and recommended raising taxes. Kennedy's economists violently disagreed. They sought instead to end the sluggish economy—not respond to its existence—

and schemed to provide the appropriate impetus by reducing unemployment. A reduction in unemployment, they maintained, could best be approached by issuing a large tax cut. Rather than raise taxes, the New Frontiersmen demanded an across-the-board reduction. The federal budget would be in deficit—for the first time during peace and without a depression. But persons would have more disposable income and buy more goods, which in turn would lead employers to hire more people to produce those goods. More people then would move on the tax rolls and federal revenue soon would more than balance that money lost by cutting taxes.

Kennedy also employed social scientists as part of his foreign policy contingent. Political scientists, psychologists, sociologists, and economists all provided theory and blueprints to enable the emerging nations—often termed *underdeveloped*—to emulate the United States and so stay out of communism's clutches. Their "modernization" theory, aimed at improving the quality of life among these often-impoverished peoples, depended heavily on exporting American scientific and technological know-how to alter traditional societies, cultures, and practices, replacing them with arrangements that had proved successful in the United States and in other Western democracies. These scholars maintained that their plans were not pipe dreams; they went to Kennedy armed with highly mathematical statistical models, predictions, and precise patterns showing that their disciplines were the products of the best and brightest.

Kennedy's grand plan for Latin America, "Alliance for Progress," was perhaps the most sustained effort of modernization. At the heart of the "alliance," which was announced on March 13, 1961, was a demand that each Latin American participant hew to a form of Western-style economic planning. Each "long-range" plan, the alliance held, must "establish targets and priorities, ensure monetary stability, establish the machinery for vital social change, stimulate private activity and initiative, and provide for a maximum national effort." It also insisted that "all the people of the hemisphere must be allowed to share in the expanding wonders of science . . . [which have provided] the tools for rapid progress." To help this transformation occur, Kennedy pledged the assistance of American colleges and universities. He also pledged the newly created Peace Corps, a government-sponsored agency comprised of young American men and women—mostly recent college or graduate school diplomats— charged by the rhetoric of the New Frontier. He volunteered their enthusiasm and expertise to help developing countries achieve material and social progress. Peace Corps volunteers worked on site, delivering intellectual and technical service to rearrange economies as well as more mundane activities such as bridge building. So that no one misunderstood the intellectual basis for the Alliance for Progress, Kennedy reminded his partners that it rested on modernization theory. "Political freedom must accompany material progress," he reminded listeners at a reception for Congress and the diplomatic corps of Latin American republics. And this freedom "must be accompanied by social change. For unless necessary social reforms" are made, "then our alliance, our revolution, our dream, and our freedom will fail."[21]

Kennedy was willing to extend the influence of federal science in other ways. His announcement of a "Consumer Bill of Rights" in 1962 assigned to government a new regulatory role. Claiming that science had unleashed an unprecedented number of foodstuffs and medicines in the previous twenty years, the president found that consumers, bombarded by marketing techniques and campaigns to boost products and lacking mechanisms to protect themselves, could not make rational choices without the aid of scientists. Protecting the public had become a federal responsibility, he argued, and he issued four rights that every consumer must have: the right to safety; the right to choose; the right to information; and the right to be heard. To make sure that consumers received the appropriate information, Kennedy stationed a person in each agency or department dealing with regulatory matters to bring the fruits of government research to consumers. He also mandated that his Council of Economic Advisers form a consumer advisory group so that the interests of consumers would be factored into economic policy. The president also increased budgets for the FDA, the Federal Aviation Agency (FAA), the Federal Power Commission (FPC), and the Federal Communications Commission (FCC), among other regulatory groups.

With enhanced budgets came new responsibilities. Kennedy called on the USDA to inspect all meat intended for the American public. In the case of the FDA, Kennedy wanted confirmation not only of safety but also of effectiveness. Food producers needed to demonstrate that the additive or drug achieved the effect that they stated; their scientists needed to provide FDA scientists with data that verified claims for the product's performance. Congress seconded Kennedy's plan and passed the Kefauver-Harris Amendment in 1962, which also required companies to report adverse events to the FDA, and to disclose the risks as well as the benefits of their products in advertisements to physicians.

His push for a clean air act moved scientific regulation to the forefront in an area where regulation barely existed. Its premise was "a greatly expanded national effort to control air pollution through research." "Criteria reflecting accurately the latest scientific knowledge" would rest at the center of the effort. Federal funds would be used to gather that knowledge and to establish regulation, not unlike the FDA's GRAS initiative. Finally, new regulations would continually be generated because the measure also sponsored the continued search for scientific data as well as state and local participation.[22]

The idea of special scientific plebiscites also found favor in other parts of Kennedy's administration. Surgeon General Luther Terry's idea of putting together such a panel to consider the new information about cigarette smoking and disease, especially heart disease and lung cancer, quickly found presidential favor. Terry polled the various professional organizations, such as the American Medical Association, and the relevant government agencies, such as the National Cancer Institute, the FDA, and the Office of Science and Technology, to develop procedures, methods, and goals for the panel. The meeting produced a list of 150 scientists and doctors especially knowledgeable on the question of respiratory ailments, excluding anyone who had lobbied for or even taken a prominent public position on the matter and allowing related agencies and organizations to reject those deemed unsuitable. By the end of 1962, Terry had his scientific commission and it began careful deliberations.

Kennedy clearly applied science to any number of problems. His infatuation with the application of scientific knowledge was matched by his appreciation of "pure" science. Yet he, like his economists, recognized that knowledge for knowledge's sake lacked social responsibility. Never has the "need been greater for the cooperation between those who work in Government and those [working] in the distant laboratories on subjects almost wholly unrelated to the problems we now face," Kennedy asserted. The reason was clear. "We realize now the progress in technology depends on progress in theory; that the most abstract investigations can lead to the most concrete results."[23]

Kennedy's "pure" science was hardly pristine as its value revolved around applicability. He heartily believed in the technological spin-offs of basic research and he emphasized basic work in the NSF and the National Institutes of Health (NIH). His support of the NSF was a matter of faith but he truly expected basic research within the NIH to produce myriad tangible remedies and understandings. Kennedy counted on increased NIH funding for the creation of new drugs such as penicillin. He also pushed established agencies in a new direction to implement his scientific agenda. For example, he worked with Congress to create an Institute of Child Health and Human Development within the NIH so as "to stimulate new interest and efforts in these research areas."[24]

THE MILITARY-INDUSTRIAL COMPLEX

Kennedy, then, used a new understanding to expand the federal roles for science. He carried federal science into new areas. But he also shepherded a massive increase in federal funding generally. Indeed, the increase was so large that even with the infusion of funding into all sorts of uncharted areas, the percentage for military/defense science funding reached unprecedented heights. In 1963 the NSF budget comprised less than 2 percent of the federal science effort. Fully 93 percent of all federal science dollars went to the Department of Defense, the Atomic Energy Commission, and NASA. Atomic weapons, nuclear submarines, portable reactors, and rocket boosters were extraordinarily expensive. And Kennedy was the ultimate cold warrior. He had run on a platform that a missile gap existed. Unwilling to cede anything to the Soviets, he moved on all fronts to confront and surpass them. In almost all cases, science funded by the federal government was at the heart of his moves.

In his inaugural address, Kennedy stated that "we will pay any price, bear any burden, meet any hardship, support any friend, oppose any foe, to assure the survival and the success of liberty." He carried this theme throughout his administration. Worried that the United States was falling behind the Soviets in nuclear weaponry and unconvinced that he possessed the surveillance capabilities necessary to monitor ongoing Soviet underground tests, he resumed atmospheric weapons testing in 1962, a practice that Eisenhower had ended some four years earlier. Alarmed by Soviet missile technology, Kennedy spared no expense on missile technology and devoted a considerable share of the defense budget to outdesign and outproduce Soviet missile makers. He also turned the ARPA's attention to the question and ordered it to establish programs to detect ballistic missile launches and nuclear explosions. Part of Kennedy's strategy revolved around the issue of preparedness. To prepare strategies in advance—to anticipate events—the Defense Department relied heavily on social scientists and others to spin out scenarios— potential situations—so that they could develop effective strategies beforehand and not be caught unaware. Among the possibilities they considered was how the United States could best deal with the reality that it would have less than fifteen minutes to respond to a surprise attack from the Soviet Union. If it did not respond within that time, it was argued, then America's missile capabilities would be destroyed. The strategy of Mutually Assured Destruction (MAD) was offered by defense analysts to resolve this predicament and Kennedy avidly applied it. In effect, the MAD scenario demanded that America increase exponentially its nuclear weapons and diversify them by location and platform so that any first strike would wipe out only a portion of the nuclear fleet. Enough would survive to destroy the Soviet Union. The threat of total annihilation—nothing left to preside over—would presumably prevent the Soviets from making a first strike.[25]

Implementing these strategies and plans required huge expenditures of federal funds. But Kennedy also braced the country for another possibility. If MAD and his other initiatives succeeded in reducing the threat of nuclear weapons by making atomic warfare unsurvivable and therefore unthinkable, then any future war would be fought with conventional weapons. Eisenhower had concentrated on nuclear capabilities, leaving the conventional armed forces in need of considerable modernization. His concern grew when Khrushchev in January 1961 gave an address titled "For New Victories of the World Communist Movement." The Soviet premier argued that nuclear weaponry made large-scale events extremely dangerous and pledged to extend the communist cause through "wars of liberation." Noting that a good part of Africa, South and Central America, and Asia were now throwing off the bonds and vestiges of Western colonialism, he recognized these liberation wars as "sacred" and promised the USSR's support in the struggle to establish communist states.[26]

Kennedy then set out to equip the army with the new challenge of fighting these smaller wars against communism. This army would not necessarily fight in Europe—although the Berlin blockade of 1961 suggested that possibility—but perhaps in the developing or underdeveloped world where there were few roads, which would make heavy weaponry a virtual impossibility, and few airports. Equipment would need to be carried in by air and smaller, mobile, easily landed vehicles might wage the air campaign. Forests would provide cover both for the new army and its enemies. Methods to detect combatants through the foliage or technology to remove foliage could prove crucial in these new battles. From these concerns came a slew of new modern weapons, applicable for new modern warfare: helicopters, all-terrain vehicles, infrared sensors, defoliants, etc. Each required extensive research and development, and each depended on federal science funds.

Kennedy's Alliance For Progress was the nonmilitary response to Khrushchev's liberation war speech; Vietnam would become a military one. The space race and Kennedy's promise to land an American on the moon by the end of the decade also stemmed from cold war competition and followed another precipitous American disappointment. On April 12, 1961, the Soviets launched a spaceship that carried the world's first cosmonaut, Yuri Gagarin. Gagarin's vessel completed a complete orbit of the planet before landing uneventfully. The launch proved fully as mesmerizing as *Sputnik* across the globe, and at least as depressing within the United States. Again, communist science and technology had proven triumphant. About three weeks later on May 5, NASA launched its own first manned success. Alan Shepard piloted a satellite 116 miles to the very beginnings of outer space just beyond the upper atmosphere, returning to the surface a scant 15 minutes, 28 seconds later.

By the end of May, Kennedy had developed a plan to recapture global attention. He called on Congress to fund his goal of landing a man on the moon by decade's end. Kennedy identified winning the space race as the crucial factor in keeping unaligned nations from joining the Soviet Union. Gagarin's flight "had captured the minds of men everywhere," Kennedy said. Victory also was essential for America's immediate security, to "not see space filled with weapons of mass destruction." For both reasons, it was critical that space not be "governed by a hostile flag of conquest, but by a banner of freedom and peace." "Space science like nuclear science and all technology has no conscience of its own," he continued. "Whether it will become a force for good or ill depends on man, and only if the United States occupies a position of preeminence can we help decide." Kennedy concluded, "Our leadership in science and industry, our hopes for peace and security, our obligations to ourselves as well as others all require us, for the good of all men . . . to become the world's leading spacefaring nation."[27]

Not all American scientists signed on immediately to the president's program. Many deemed the program not particularly scientific. Little science would be undertaken. It was simply technique. Others claimed that more imminent problems existed on earth for science to tackle. A few countered that machinery rather than men should explore space. NASA recognized the dissention and worked hard to develop broad-based scientific support. It contended that the program would spin off all sorts of technological marvels and more than pay for itself. It further argued that the moon quest would lead a generation to increase support for science generally, support that could be channeled into other projects in other areas. NASA also held conclaves of scientists to press its viewpoint. It held a two-month long conference for one hundred scientists in Iowa at the cost of more than $500,000. Not surprisingly, the gathering concluded with a resolution praising the moon effort.

John Glenn's 1962 orbital spaceflight cheered American prospects. But NASA scientists realized that the rocket booster used on the flight was much too limited to be of future use. NASA approached the booster by following now-established procedures: It created competing programs, one to develop a solid fuel booster, the other a liquid fuel one. The more powerful and dependable—NASA scientists estimated that it would require a device at least forty times more powerful than Glenn's rocket—would carry man to the moon. Just in case, it also stepped up work on the Rover rocket, a nuclear-powered booster. Yet the agency recognized public support as a key component to its success, and in the year after Kennedy first announced the moon program it launched more than thirty satellites, including those for communications and weather observation. *Telstar* was especially noteworthy. A joint AT&T–NASA project, the combine launched *Telstar* on July 10, 1962, as the first of several satellites positioned to provide worldwide communication. To celebrate its launch, *Telstar* beamed a color television signal to the United States from Europe, the first live transatlantic video broadcast. The picture was shown across the country. NASA also launched satellites instrumental to the national defense. Its Transit series helped America's nuclear submarines navigate successfully. Its Midas satellites used infrared sensors to detect hostile missile launches. NASA's Samos orbitals provided crucial global military reconnaissance. Despite these and other military applications, Kennedy continued to maintain that the peaceful use of space marked a fundamental difference between the United States and the Soviet Union. He also argued that the US satellites were better, proudly proclaiming that those launched during his administration "were far more sophisticated and supplied far more knowledge to the people

of the world than those of the Soviet Union." In 1963 NASA operated with a $5 billion annual budget, more than seven times the
budget it had under Eisenhower.[28]

Kennedy clearly recognized that his dramatic expansion of governmental science required an equally impressive number of scientists. To ensure that this would occur, he prevailed upon Congress to
extend and expand the National Defense Education Act. In the modification, he made sure that more money per year was available and
that those science schools in greatest demand were able to support
a higher proportion of students. He also convinced Congress to
enact legislation channeling $1.2 billion to universities for "class-
rooms, laboratories, and libraries," places to train the next generation of leaders. But perhaps his biggest triumph for regularizing production of scientists and science at universities was getting Congress
to approve the regular payment of indirect costs associated with scientific work. Done according to a fixed formula, the federal government now paid not just for the scientists, equipment, and the like,
but universities were entitled to moneys for use of their buildings,
electricity, administration, and sundries for those scientists receiving
federal funding. This overhead money in turn could be employed to
further the institutions' scientific infrastructure, making future
grants more likely.[29]

By then Kennedy had completed a transformation in American
governance. After his administration, there could be no ambivalence;
there was no turning back. Science was now thoroughly intertwined
within the national fiber. Through NASA, the Defense Department,
and the Atomic Energy Commission, Kennedy oversaw a military-
industrial complex that far surpassed anything Eisenhower had envisioned. These departments, coupled with Kennedy's vision that
science should answer America's social questions, depended on universities, their staffs, and students for a significant portion of the
scientists and a great part of the science. As a consequence, the federal-
university reciprocity became much deeper than ever before. Science
achieved a new prominence on university campuses that depended on
and was paid for by federal government dollars. The NDEA was not
only continued but also its programs were expanded, making it easier

to go to college and more necessary to do so. The number of physical scientists receiving degrees doubled every fifteen months. As the dollars grew and as more scientists, more institutions, and more disciplines became conditioned to feeding at the federal trough, their dependence on federal largesse intensified and their expectations increased. Universities factored in federal research moneys as parts of their recurring budgets, anticipating that their scientists would draw an increasing amount of dollars yearly.

Scientists overwhelmingly approved of this arrangement. Kennedy's broadening of federal initiatives beyond the physical sciences brought even more researchers and disciplines into the fold. But this research money was neatly circumscribed; it went to fight the cold war, broadly conceived, and to approach America's social problems in a manner that was congruent with the political goals of the Kennedy administration. In effect, the overwhelming portion of federal science was targeted science where money was available to undertake specific tasks. Scientists had a choice, at least in theory: they could investigate those social and/or political questions or forego federal money. But scientists felt pressured to continue accruing financial support, and increasingly, only the federal government offered that kind of funding.

The public seemed even more enthusiastic than the scientists. In the cold war world, Americans' freedoms, choices, diplomatic partners, and perhaps even survival seemed to depend on scientists and their productions. As important, science appeared as the basis of the future; it would yield new foods, new medicines, new understandings of the human condition, new substances, new drugs. Having the federal government oversee private enterprise seemed like a good idea and its increasing use for scientists and scientific measures and methods apparently provided a potent sense of well-being. Science would secure consumer interests even as it maintained America's military security.

CHAPTER 2

COMING APART

THE STATUS QUO?

Kennedy's assassination on November 22, 1963, gripped the wary nation. Lyndon Baines Johnson, Kennedy's vice president, assumed office and almost immediately began to exert himself. While more experienced within the Executive Office than any previous twentieth-century vice president, he had not been a member of Kennedy's inner circle and his course as president largely remained a question. Certainly his persona was considerably different. Johnson was coarse and folksy, not urbane and handsome. He was straightforward, not wry and witty. Kennedy was inspiring. Johnson worked behind the scenes. But no one doubted Johnson's experience or savvy. He had served as a most effective Senate majority leader before he accepted the vice presidency. He was a master at getting people to accept his agenda as he used his fabled "treatment, an incredibly potent mixture of persuasion, badgering, flattery, threats, and reminders of past favors and future advantages" to get his way. The relentless Johnson remained true to Kennedy's goals and ambitions and, for starters, called on Congress to pass legislation favored by the martyred president as a tribute to his memory. But Johnson did much more than that. He used the nation's goodwill and his political skills to goad, cajole, and demand that Congress follow his lead.[1]

To the vast majority of scientists and engineers, Johnson's pledge to enact Kennedy's program signaled that the new president shared his predecessor's affection for and reverence of science and that the new administration would follow closely in Kennedy's footsteps. Johnson gave signs that this would be the case.

Although he announced his intention to resign well before Kennedy's death, Jerome Wiesner agreed to continue for a time as science advisor and reassured the scientific community by seeming to have the new president's ear. The man who succeeded Wiesner, Donald F. Hornig, had worked in science policy for Eisenhower and Kennedy. In fact, Kennedy had already tapped Hornig to succeed Wiesner, and the new president approved. Johnson, who as vice president had chaired the National Aeronautics and Space Council, remained fully committed to landing a human on the moon by the end of the 1960s. As if to highlight the significance of the moon landing, he renamed Cape Canaveral, NASA's operations center, the Kennedy Space Center. Johnson realized both the symbolic and military value of the space program and continued to fully fund the expensive enterprise. In the months after the Kennedy assassination, NASA built, tested, and launched the initial version of its Gemini spacecraft. (McDonnell, a staunch member of what Eisenhower had termed the military-industrial complex, was the contractor.) Although Gemini was designed to hold two astronauts, the first launch was unmanned to ensure that the launch vehicle and capsule could withstand the rigors of spaceflight. The Gemini program was to prepare astronauts for the eventual moon landing. Its goals included training astronauts to work in space for a protracted period, rendezvous and dock with another vehicle, perfect methods

http://history.nasa.gov/ap11ann/kippsphotos/5875.jpg

Fig. 4

After President John F. Kennedy committed the United States space program to the goal of landing a man on the moon and returning him safely to the earth, the National Aeronautics and Space Administration invested millions in solving the technical, scientific, and practical problems involved. Culminating a series of important missions, *Apollo 11* astronauts landed on the moon on July 20, 1969. Mission commander Neil Armstrong descended to the surface while making the comment that became famous, "That's one small step for a man, one giant leap for mankind." Armstrong then set up equipment on the moon for several scientific experiments along with astronaut Edwin "Buzz" Aldrin, shown here posing with the American flag that the two placed on the moon.

of reentry, and collect information on the effects of weightlessness during extended flights.

In 1964 NASA also worked on the Saturn I rocket, a version of the vehicle scheduled to take men to the moon, and launched more than a handful of satellites. Primarily designed to measure the weather, meteor showers, and the geography of earth, these satellites were much bigger and more complex than those sent into orbit under Kennedy. But perhaps the most optimistic act that NASA participated in during the first year under Johnson came at Johnson's behest. Aware that ventures into space required long planning and much money, the new president urged NASA's administrator to begin thinking about what challenge the space agency wished to tackle after it completed its moon landing. (See fig. 4.)

Johnson wanted NASA to establish its next great national project even as it was just beginning its current one. That sense of planning was common during Johnson's first year. The president set for the country a broad, ambitious agenda for the future. But few of these new national initiatives would be directed at the military or defense. While he upgraded the antiballistic missile defense—nuclear-tipped rockets that could explode high in the atmosphere and disable incoming enemy missiles—Johnson claimed that Kennedy's three-year military spending spree had fully modernized America's weaponry. Indeed, by 1964 the cold war seemed less an immediate threat, more a chronic than an acute condition. Coupled with the recent nuclear test ban treaty, which outlawed atmospheric nuclear tests, and in wake of the Cuban missile crisis and the Bay of Pigs, Johnson proposed a slight cut in defense spending and pledged to move toward further nuclear arms control agreements. In fact, as a first step and to show good will, he promised that the United States would "not stockpile arms beyond our needs or seek an excess of military power that could be provocative as well as wasteful." To that end, he cut the production of enriched uranium by a quarter and decommissioned four plutonium piles, thereby decreasing markedly the material necessary to make additional nuclear devices.[2]

Johnson's production cut of nuclear weapons–grade material demonstrated a commitment to the control of nuclear arms, a stance applauded by many scientists who worried about the bomb's awful power. Johnson's enthusiastic endorsement of his predecessor's support of nuclear power also cheered a vast majority of the science community. Kennedy had appointed the Nobel Prize–winning chemist Glenn Seaborg to head the Atomic Energy Commission and moved forcefully to design nuclear power plants to supply America's energy needs. In the first year of Johnson's watch, a new kind of nuclear reactor went on line as scientists at national laboratories tried to determine the relative merits of water and liquid metal–cooled devices.

Fulfilling the nation's energy requirements through nuclear reactors was in some ways similar to the quest to land a man on the moon by decade's end. Both programs embraced new technology and sought to apply it in a way to impress and inspire. Johnson supported several other national initiatives during his first year. Kennedy had initially proposed most of them and all incorporated science and scientists at their heart. But unlike Kennedy, the ultimate cold warrior, Johnson focused extensively on domestic policy, even proposing a modest decrease in foreign aid. He argued that what happened domestically was at least as important for keeping the free world free as was fighting the communist menace. As the whole world watched America, he called on Congress to refashion society in such a way as to convince others definitively of democracy's superiority. We have now, he argued, "a unique opportunity and obligation to prove the success of our system; to disprove those cynics and critics at home and abroad who question our purpose and our competence." In his initial State of the Union message less than two months after Kennedy's death, Johnson threw down the gauntlet by declaring an "all-out war on human poverty and unemployment in these United States." He brazenly dared Congress to pass legislation to root out and end what he saw as the endless cycle of poverty. He demanded that individuals receive "a fair chance" to escape impoverishment through "better schools, and better health, and better homes, and better training, and better job opportunities." He challenged Congress to "finally recognize the health needs of our older citizens," to reform "our tangled transportation and transit policies," to achieve "the most effective, efficient foreign aid programs ever," and to help "build more homes, more schools, more libraries, and more hospitals than any" other Congress "in the history of our Republic."[3]

In many ways, Johnson's War on Poverty attacked the issue with the same gusto and zeal that previous administrations had approached defense- and military-related questions. Like his predecessors, Johnson harkened back to the glory years of World War II when failure was unthinkable and national quests were the order of the day. In particular, his poverty campaign took inspiration from the Manhattan Project, where teams of scientific and technical experts under intense pressure designed and produced what seemed the ultimate weapon. There even the inconceivable was conceivable, an approach Johnson brought to bear on poverty. "I have no illusions," he contended. While he did not expect poverty to end "in my lifetime," Johnson did expect programs, scientifically constructed and applied, to "minimize it, moderate it, and in time eliminate it." Johnson reserved a prominent place in this war for himself, of course, far more important than simply galvanizing public attention. He arrogantly considered himself—rightly so—a master at translating vision into legislation and legislation into law. While in the Executive Office, he would propose legislation and methods and then attempt to act as he had when he served as Senate majority leader. He would repeatedly use the "treatment" to get the plans framed by his experts enacted by Congress.[4]

Four months after declaring war on poverty, Johnson was articulating an even grander program. Demonstrating an eloquence and passion few believed he could express in public, Johnson challenged younger Americans to use "your imagination, your initiative, and your indignation" to "enrich and elevate our national life." Make "progress . . . the servant of our needs," thereby transforming our already "rich society and powerful society" into a "Great Society." The Great Society, he continued, "is a place where every child can find knowledge to enrich his mind and to enlarge his talents." Johnson identified "in our cities, in our countryside, and in our classrooms" as places to begin constructing the Great Society. Johnson then went on to recount a litany of problems that would affect this and subsequent generations: decay of inner cities and despoiling of suburbs; poor transportation routes and facilities; threats to "the water we drink, the food we eat, the very air that we breathe;" and disappearing green spaces, overburdened seashores, and overcrowded parks. But he was most precise about what was necessary in the classroom. Johnson argued that one quarter of the nation's adults never finished high school; that every year more than 100,000 high school graduates could not afford college while noting a projected increase of 5 million school children over the next decade; that many public schools and colleges were overcrowded and out of date; and that a good proportion of the nation's teachers were ill-prepared to lead the next generation.

Johnson understood the magnitude of the challenges. Therefore, he promised "to assemble the best thought and the broadest knowledge from all over the world to find those answers for America." Using "working groups" he would hold a series of White House meetings. "From these meetings . . . and from these studies we will begin to set our course toward the Great Society."[5]

During 1964 Johnson created at least fifteen commissions or task forces to study and suggest methods to achieve the goals articulated in his Great Society and State of the Union speeches. Each of these task forces, whether investigating some aspects of urban renewal, trade, transportation, or agriculture, was similar in that it was comprised of professionals most involved in the question at hand. Like Kennedy, Johnson recognized the concept of expertise and turned to leaders in the field to engineer real-world solutions to problems. He apparently understood that those long active and involved in the field realized the hurdles, obstacles, and opportunities to surmount problems. What they had lacked was federal power, money, and clout, and Johnson stood ready to enable these men and women to ply their various trades, professions, and expertise. As important as formulating propositions, the ever-political Johnson recognized the virtue of professional assessment. Suggestions by scientists seemed to hold the same reverence that the methods of science held. If the scientific method was impartial, nonpartisan, and indisputable, then scientists were also. As it had in the earlier era, vast segments of the populace and to a lesser extent Congress deferred to scientists because of their purported expertise, and Johnson realized he could use the patina of science to help ram measures through Congress.

In Johnson's Great Society, scientists would refashion American society to present its benefits to all and to protect those things that made life worth living. But as Johnson articulated this vision, he reached out to and depended on groups of scientists not involved in defense, space, or the nuclear power industry, the traditional recipients of federal funding. His domestic agenda would incorporate social scientists, engineers, and physicians rather than physicists and, to a much lesser extent, chemists. These newly consecrated men and women brought their expertise to bear on specific problems and applied it. They did not chomp at the bit to do "pure" research, confident that some day their insights might lead to some social solution. Quite the opposite was the case. Like Kennedy's economists, Johnson's social scientists recognized society as their laboratory. Theories not applied were simply idle speculation, not subject to rigorous circumstance and observation. In effect, Johnson tapped scientists that had not been fully integrated into the concept of federal largesse and brought them in as an indispensable part of governance. At the same time, he turned his attention from those disciplines organized around the idea of constant, dependable, increasing federal money where oversight was somewhat distant.

Among the first issues that Johnson tackled was health. He did not initiate the first salvo but benefited from a measure Kennedy had put in plan years earlier. On January 11, 1964, Surgeon General Terry's Advisory Committee on Smoking and Health issued its final report. Meeting in the same room that handled news conferences after Kennedy's assassination to connote the event's importance, the committee released its findings on a Saturday to prevent a run on tobacco stocks. It implicated cigarette smoking in several diseases and identified it as "a health hazard of sufficient importance in the United States to warrant appropriate remedial action." Although no remedial action was even proposed in 1964, the surgeon general established later that year the National Clearinghouse for Smoking and Health to gather information about the health risks of cigarettes and to induce persons to quit smoking.[6]

Johnson's health crusade picked up steam on February 10, 1964, when he outlined to Congress a broad, comprehensive program to improve the nation's health. Reminding Congress that the spectacular success of modern medicine came at a cost, Johnson noted that when people live longer, they become liable to various degenerative maladies. Medical progress had come at increased research costs, and the rising cost of medical care reflected those increases. America's population had grown by more than 25 percent in the previous fifteen years, straining an already tight supply of medical personnel. Technological progress often came at an environmental cost, polluting the air, water, and food.

Perhaps as a harbinger of the Great Society, Johnson declared first-rate medicine a right of all Americans, not a privilege of some. Everyone, no matter their financial status, was entitled to the fruits of modern medicine. Thousands suffered from preventable diseases; since he recognized inadequate healthcare as a leading cause of poverty, he called broad-based access to modern medicine a critical element in the War on Poverty.

Johnson proposed sweeping new health initiatives. To improve the quality of care and to expand access, he urged Congress to enact federally financed hospital insurance tied to social security for the elderly, to use federal money to expand hospital construction in areas not adequately served, and to modernize long-standing, now outdated hospitals and fund the establishment of long-term care facilities. He next focused on mechanisms to improve the quality or increase the quantity of caregivers. Congress needed to finance creation of new schools of nursing as well as expansion of established programs and to enact special scholarship and loan programs to assist people to become nurses. Congress must enact similar legislation to assist public health workers and the colleges that trained them. Caregivers for the mentally ill and mentally retarded required federally sponsored training to assist those afflicted after deinstitutionalization.

Johnson also concentrated on enforcement as he recognized the cost effectiveness of prevention, and he labored to have Congress forcefully legislate in that direction. Specifically he wanted appropriations in order to develop plans to establish new environmental health facilities, to hire additional FDA scientific and regulatory personnel, and to sponsor research about the effect of pesticides in the environment. He also wanted pesticides scrutinized in a manner similar to foods. Johnson sought passage of legislation that outlawed sale of pesticides until "a positive find of safety" and he wanted to mandate cost-benefit analyses of pesticides with respect to human health, domestic animals, and wildlife before any chemical agent could legally be used in agriculture.

In addition to these programs, Johnson mentioned three related task forces or commissions. Two of them, the Presidential Advisory Commission on Narcotics and Drugs and the Task Force on Research Funding in the National Institutes of Health, were straightforward. His Presidential Commission on Heart Disease, Cancer, and Stroke was a grander venture. Kennedy had convened a White House conference to address those questions but with no result. Johnson was more optimistic. Having witnessed the conquest of polio and discovery of antibiotics, the so-called wonder drugs, he thought it possible for men and women to live regularly to 100. The three diseases accounted for 71 percent of deaths in America and more than $40 billion in yearly health costs. To Johnson, much of the medical science necessary to extend life already existed. What was missing was a means to regularize its production and to ensure its dissemination throughout the American population.

To that end, Johnson tapped the noted heart surgeon Michael DeBakey to chair the commission. Made up mainly of prominent physicians and life scientists affiliated with research facilities and fortified by persons active in medical philanthropy, the committee was greeted enthusiastically. While a publication or two wondered if America could afford to conquer disease, the vast majority argued that the nation could not afford not to try. Commentator after commentator compared the quest for a moon landing to the quest for health and asked why the moon landing received billions and health merely millions.

ELECTION OF 1964: POLITICAL RUBICON?

The DeBakey committee began deliberations almost immediately. Its origins coincided with the onset of the 1964 presidential campaign. The Republican nominee, Barry M. Goldwater, a former military man and senator from Arizona, won his party's endorsement. In sharp contrast to Johnson's Great Society, Goldwater objected to extensive federal intervention in domestic affairs, claiming that unchecked government power "leads first to conformity and then to despotism." It also killed initiative. America under Johnson, claimed Goldwater, was "plodding at a pace set by centralized planning, red tape, rules without responsibility, and regimentation without recourse." Rather than making "people conform in computer regimented sameness,"

Goldwater cherished "diversity of ways, diversity of thoughts, of motives and accomplishments." To Goldwater, government existed to "secure . . . rights and guarantee . . . opportunity," which limited its functions to law enforcement, maintaining economic stability, and the national defense. And it was this last aspect of governmental action where Goldwater found Johnson sorely lacking. He let the nation's armed forces slip after Kennedy's buildup, a major mistake according to Goldwater since he demanded that America confront communism more aggressively. In particular, the Republican nominee wanted NATO commanders in Europe to have the opportunity to use tactical nuclear weapons without approval from Washington. More generally, he vehemently disagreed with attempts at disarmament and bans on nuclear testing. America alone was the arbiter of when the nation should act militarily in the interests of democracy and if that meant using nuclear weapons, Goldwater favored such unilateral moves. It is time, he said, for "thoughtful men [to] contemplate the flowering of an Atlantic civilization, the whole world of Europe unified and free, trading openly across its borders, communicating openly across the world." This, he concluded in direct repudiation of Johnson, "is a goal far, far more meaningful than a moon shot." In fact, Goldwater demanded that NASA and its civilian scientists be terminated and all space activities vested in the military.[7]

The politically savvy Johnson and his supporters quickly capitalized on Goldwater's prescription for the future. Led by Donald M. MacArthur, a physical chemist, the husband of Lady Bird Johnson's niece, and later deputy director of research and engineering in the Defense Department, and joined by the recently resigned Wiesner, other presidential science advisors, and many members of Johnson-appointed task forces or commissions—including DeBakey, these partisans formed themselves into a political entity to support Johnson's reelection. Finally named Scientists, Engineers and Physicians for Johnson–Humphrey, the group sought to mobilize scientists, engineers, and physicians to openly and actively back the Democratic ticket. This was no small-scale effort. They met for two days in August and immediately thereafter established "grassroots" organizing committees to enlist scientists in each state to the cause and to stump for the Democratic ticket in that state. Making speeches, going door-to-door, phoning voters, and stuffing envelopes, more than 100,000 scientists joined the organization prior to election day.

This unprecedented and dramatic move into politics by a huge number of active scientists demonstrated just how much antagonism Goldwater's platform engendered. To be sure, his militarism and stance on nuclear weapons horrified many scientists and citizens. Johnson's campaign played on those fears by producing several television commercials to expose possible consequences of electing Goldwater. One centered on the bucolic walk of a pregnant woman and her daughter while a female voiceover, accompanied by the appropriate music, warned that such an amble would not be safe had it not been for the nuclear test ban, which removed fallout-produced strontium 90 and cesium 137 from polluting the atmosphere. Another pictured a young girl of about five years of age lovingly licking her ice cream cone. There the motherly voiceover mentions that the test ban makes such a delight possible but warns that one person, Barry Goldwater, wanted to abrogate the test ban treaty and if elected, might well start testing again. But the most memorable focused on a little girl playing with a daisy. Counting as she pulls off petals, she soon gets confused and starts a downward count. At that point a harsh, almost robotic voiceover starts a countdown as the camera focuses ever closer on the girl's eye. When the countdown reaches zero, the girl's pupil is replaced by an ever-expanding mushroom cloud at which time Johnson's voice tells us "these are the stakes . . . We must love each other or die." As the commercial fades to black, a voiceover ends the commercial with a refrain common to all Johnson commercials. "The stakes are too high for you to stay home."[8]

Indoctrinated in a professional culture in which participating in atomic bomb development seemed to raise a series of moral issues, many scientists took the use of nuclear weapons personally and demanded that every option be explored before (if ever) using nuclear options. Goldwater's approach appeared to them cavalier, foolhardy, and idiotic. Scientists derided what one critic called "Goldwater's mixture of the big stick, the large mouth, and the small brain." Yet that politicization of science—creation of atomic weaponry and subsequent interactions with America's defense establishment—did not dissuade them from reentering the arena in the face of Goldwater's statements and nomination.[9]

But nuclear annihilation was not the only threat that scientists recognized from Goldwater. They also understood that nuclear war was bad for the science business. Now heavily dependent on federal funding, war would not only take away various revenue streams but it would also compel scientists to maintain their research programs by moving into other less desirable areas, ones that they had not selected. Yet that was not the half of it. Even with no nuclear event or change, scientists overwhelmingly found Goldwater's platform a catastrophic disaster. The issue was his view of the federal government. By demanding that the federal government remove itself from most of its activities, scientists knew that this would result in a precipitous and calamitous drop in scientific research funding but also in the status and public opinion of scientists. In short, Goldwater's vision would end science as they had come to know it. One prominent scientist succinctly summed up the case. "With his lack of education, his know-nothingism, and his nostalgia for a past that never was," Goldwater would "foreclose the future for all of us."

The perceptive and cagey Wiesner never missed an opportunity to remind scientists that Goldwater's election would undoubtedly kill science in the United States. As important, he rarely seemed partisan when making the claim, always arguing that for the past three decades presidents and leaders from both parties, as well as the overwhelming majority of the citizenry, favored the massive extension of the scientific enterprise and the federal government's primary responsibility for having created, fostered, and financed it. Johnson's reelection, he asserted, was the only way "to continue the dramatic economic and social developments of the past twenty years." Johnson, Wiesner continued, "like President Kennedy before him, has a keen appreciation of the value of education, scientific research, and technology in our society. He believes in and supports the programs created and expanded by Presidents of both parties during the last thirty years." Well aware that many in his audiences held science posts at colleges and universities, Wiesner often reminded those folks that "government support for research in the universities is one of the prime reasons for their greatness in science and engineering."

Wiesner's portrait of his boss did not always go unchallenged. Scientists often found Johnson distant, aloof, less enamored with scientists than Kennedy had been. Rather than a person who treasured scientists for their minds and creativity, Johnson seemed to appreciate scientists because of what they could do to further his agenda. Wiesner confronted this question, although not exactly straight on. "I have heard people call President Johnson a politician, with the implication that to be a politician is bad," he acknowledged. Rather than a vice, Wiesner claimed it a virtue, a skill that would enable Johnson to move his agenda, which appeared quite favorable to scientists, through Congress. "The President must be an effective politician if he is to promote the country's programs," Wiesner maintained. Johnson exhibited "skill and confidence," masterfully reconciling "conflicting views in order to create his program and then to move it through the Congress." In short, Wiesner told scientists that Johnson deserved their backing because he favored extensive financial support for science and possessed the political skills necessary to translate that belief into public policy.[10]

A Great Society?

Johnson won an unprecedented electoral victory. For the second consecutive year, Johnson used his political abilities to force his will on sometimes-reticent congressmen and to get his measures passed through Congress. In addition to epic civil rights legislation, he gave congressmen the "treatment" in a wide variety of areas. Almost all extended the range of the federal government, sought to achieve some modicum of social justice, and placed scientists and scientific methods in the forefront. In the field of regulatory science, for example, Congress at Johnson's insistence established the government's first clear-water purity standards and a Water Pollution Control Administration to make sure that states were indeed keeping their waters pure. This legislation also empowered the federal government to assume jurisdiction over state waters if a state failed to act adequately. To help states achieve the desired result, the law provided $150 million yearly to construct new waste and water facilities and established a grants program of $20 million yearly to engage in research and development of sewage treatment methods.

Johnson and Congress also tackled other Great Society and War on Poverty issues. For the first time, federal regulators—scientists and technicians—faced the obligation to set motor vehicle emission standards for carbon monoxide, hydrocarbons, and other pollutants. Manufacturers received fines for not meeting them. Following up on Surgeon General Terry's *Report on Smoking and Health*, Congress mandated warning labels on all cigarette packs and cartons, warning CAUTION: CIGARETTE SMOKING MAY BE HAZARDOUS TO YOUR HEALTH. Congress formed the executive Department of Housing and Urban Development (HUD), to oversee federal loan programs but also to investigate all issues of community development and to translate that research into action. One of its first undertakings—cosponsored by the Office of Science and Technology—was to invite sociologists, psychologists, architects, city planners, engineers, and other scientists to a days-long White House Conference of Science and the City. By gathering groups of scientists together, HUD hoped to draft plans for a livable, vital urban, and urbane future.[11]

Education received special attention. As a means to break what Congress and Johnson recognized as the cycle of poverty and in response to a Johnson-called task force of educational scientists, the Office of Economic Opportunity established a pilot program to take disadvantaged youngsters at age three and prepare them for public schooling. Project Head Start made it the Office of Economic Opportunity's responsibility to get these programs established and offered grants to individuals and cities to erect Head Start centers. Another task force focused on the federal government's responsibility to support research on education. It called for a number of national laboratories, similar in spirit to the national laboratories of the Atomic Energy Commission, to produce a "system . . . for continuous [educational] renewal." Through close linkages among schools, state departments of education, and research universities, these grand national laboratories "would develop and disseminate ideas and programs for improving educational practices throughout the country."[12]

Title IV of the Elementary and Secondary Education Act was a result. In 1960 the Office of Education had a budget of less than a half-billion dollars. By 1967 the budget had risen to nearly $4 billion. The budget for the laboratories alone was more than $100 million. A National Center for Educational Statistics operated within these new laboratories to ensure that educational scientists had the most current and correct information upon which to do their research. In a further effort to pattern education after more traditional scientific endeavors, it created the Educational Resources Information Center (ERIC). Modeled after the NSF's Clearinghouse for Federal Scientific and Technical Information, this division was charged with collecting, disseminating, and preserving cutting-edge and exemplary research, and lessons produced by that research.

Higher education also gained Congress's and Johnson's attention and significant new resources. Here the emphasis rested on making college and university education accessible to persons of middle class and less means by providing money for students in those categories. By making college education more affordable and increasing the number of matriculated students, the federal government ironically encouraged institutions to increase tuition, a move that allowed them to expand the number of faculty, laboratories, and other facilities to meet the baby boom generation. By continuing and expanding provisions of the NDEA (1958), by creating new educational grants for the economically disadvantaged, by subsidizing interest-free and reduced-interest loans, and by establishing work-study programs to funnel funds to colleges so they could hire students to perform tasks while earning spending money, the federal government in effect transferred about $4.5 billion from the federal treasury to colleges and universities. The schools then quickly raised tuition and fees to capitalize even further on these bountiful new revenue streams. Tuition at many schools doubled in two or three years.

This massive infusion of pass-through money plus other programs suited to urgent national needs established at universities community service programs "to assist in the solution of community problems in rural, urban, or suburban areas." Recognizing that libraries were the laboratories of social scientists and humanists, the act provided $150 million to universities to acquire periodicals, books, and other material for their libraries and an additional $45 million for them to establish or enhance programs in library science. Librarians so trained pursued the acquisition and cataloguing of knowledge scientifically, standardizing the material so scholars could access it systematically. Finally, the act furnished moneys to train master teachers, to establish an apprentice teacher system, and provided nearly $150 million to improve university classroom instruction, which in practice meant scientific equipment to teach principles in physics and chemistry lectures and laboratories.[13]

Health also received its due as Congress passed a number of critical pieces of legislation. Virtually everything that Johnson had proposed in his health message in early 1964, including Medicare, was ratified by Congress the next year. In 1964–65 Congress enacted more than forty medical bills, more than all the Congresses together since 1789. But that marked the extent of the Great Society as envisioned by Johnson. What befell the recommendations of his special task force on heart disease, cancer, and strokes proved emblematic of the new difficulties facing the administration as it sought to apply science to social problems.

The DeBakey committee's report in December 1964 could have been written by Johnson. It mimicked the other measures proposed by the Johnson administration as it fought its War on Poverty and attempted to create a Great Society. The several-hundred-page report listed more than eighty recommendations for government action that would revolutionize the treatment of these three diseases in America. At its heart rested the assumption that access to healthcare should be extended to the entire population and that such healthcare must be the best conceivable care. In the spirit of other Johnsonian programs, the committee did not simply want healthcare for the masses. Like Johnson himself, the group demanded that the masses receive the same quality healthcare that they personally received. To Johnson and his task force, nothing was impossible in a Great Society. It had no limits and its fruits should be spread equally throughout the nation. This sentiment was not restricted to race, age, gender, or class.

Geographic equality also was a tenet of the DeBakey report. Persons in rural Nebraska deserved the same quality of services as those in New York City or San Francisco. The group recommended establishment of a sixty-node network of regional centers—twenty-five for heart disease, twenty for cancer, and fifteen for stroke. These nodes would be sprinkled throughout the nation, taking advantage of established universities, medical schools, and hospitals fully empowered to build and staff new facilities wherever the necessary institutional infrastructure was lacking to ensure equal access to the populace. Federal funds would pay for the medical care of the indigent. The methods suggested by the committee also reflected other Johnson programs, each of which was built upon research, access, and equitable nationwide distribution. These widely dispersed regional centers "would be strongly oriented toward clinical investigation and fundamental research. They would conduct training programs for personnel" to deliver quality healthcare. Each would be allied with a hospital as well as outpatient facilities. The staffing of each of these facilities was large and complex. In addition to the technical staff—nurses, laboratory assistants, technicians, and the like— each heart disease center, for example, would employ at the minimum "internists, cardiopulmonary physiologists, cardiologists, peripheral vascular specialists, cardiac and vascular surgeons, biochemists, statisticians, epidemiologists, radiologists, and geneticists."

Medical colleges were of central concern in this regional center scheme. Clearly the committee felt that medical colleges were the single most important structure and as a consequence embraced a program to create "centers of excellence" throughout the country. Medical schools lacking in resources, facilities, equipment, and faculty could become recipients of this government-funded program and build themselves up to the quality and competency of the most elite institutions in America.[14]

Knowledge produced at these new centers of excellence as well as the other regional centers would be transferred to other medical personnel who in turn would staff 550 new diagnostic and treatment stations. Again, segregated by disease and placed to provide similar levels of quality care throughout the states, these stations would deliver both emergency and routine treatment, and do significant screening to prevent the onset of new disease. These units were also scheduled to acquaint local physicians with the newest theories and to train them in the latest techniques for treating each disease, and they would teach the citizenry to reduce their susceptibility to it.

Public response to the committee's proposals was overwhelmingly positive and the Johnson administration quickly turned its recommendations into a bill, which it submitted to Congress. To quell possible resistance to the bill by healthcare professionals, already upset by Medicare and fears of "socialized medicine," the measure included a stipulation that the regional medical center system would not disrupt "existing patterns or financing of patient care, professional practice, or hospital administration." The bill sailed through the Senate but met considerable objection in the House. There the American Medical Association (AMA) proved more powerful and more organized. Its members complained that establishment of federal research centers and hospitals would lead to a series of carpetbagging physicians descending on rural areas and usurping longtime family practitioners. Others felt insulted by the measure's presumption that disease rates were so high because physicians failed to employ the most modern knowledge and to use the newest techniques and treatments. Several meetings between the AMA and White House staffers failed to find common ground. Ultimately, Johnson intervened and compromised to get something passed. The result was the Heart Disease, Cancer, and Stroke Amendments of 1965.[15]

The amendments proved a pale reflection of the committee's recommendations and the bill that the Johnson team had introduced into the Senate. The duration of the measure was reduced from five to three years and appropriations fell from $1.2 billion yearly to $340 million. The act no longer spoke of a full-fledged, integrated, nationwide program but rather pilot projects, and allowed for no new construction and no diagnostic stations. Worse still, it did away with regional centers as federally controlled and replaced them with "regional cooperative agreements," each of which received federal funds but retained local control and maintained traditional financing of patient care.

This was a huge repudiation of the president's plan. DeBakey among others felt betrayed by Johnson's willingness to compromise. Such a massive failure of Johnson's ability to work his "treatment" on Congress and the AMA seemed virtually unthinkable just a year earlier. What was different? What had changed?

In several ways, the political landscape in which Johnson operated in late 1965 was remarkably different than it had been just a year earlier. Certainly the political composition had shifted markedly to the Democrats, who now dominated both houses. But with one-party rule came an unexpected by-product. Freed from the idea that party discipline was necessary to ward off Republicans, Democrats began to repudiate their president. Johnson himself had opened the door to these tactics. Unlike his predecessors, President Johnson acted as if he were also majority leader. More than any president, he ignored the idea of separation of powers and not only framed legislation but also worked tirelessly and full-time to force it through Congress; his interest in the actions of Capitol Hill and his demands that they enact his programs without delay rankled many. His "treatment" made congressmen bristle; and when the threat of an alternative party decreased, Democratic congressmen took the opportunity to chasten the president and put him back in his place. He had run roughshod over too many committee chairs, seized too much power, and offended too many local congressional constituencies. Congressional hearings, formerly conducted with decorum, became contentious and hostile when White House staffers were witnesses. Newly elected congressmen lacked a history with Johnson and were committed to the folks who got them elected. With all politics being local, they felt justified in taking presidential initiatives and reshaping them to fit their own purposes and to serve their constituencies, not the ambitions of the president. Also, the new Democratic supermajority was less cohesive than it might have appeared. Southern Democrats not only objected to much civil rights legislation but they also rejected the idea that the federal government be the arbiter of the nation's social conscience while acting as its planning, implementation, and enforcement arm. With the Republican menace absent, Southern Dems often felt free to act on their objections.

Technocracy Unleashed

Johnson's influence in Congress waned generally but that stood as only one of the many difficulties facing the sitting president. Increasingly during 1965, scientists expressed dissatisfaction with and exasperation at him. This was ironic because the president had declared 1965 "the Year of Science." The matter that immediately caused a ruckus was the aftermath of the 1964 election. That more than 100,000 scientists, engineers, and physicians had joined the political fight and signed on to back one candidate gave them a sense of pride and, when Johnson won, a potent sense of entitlement. Science deserved rewards for its support. And what science wanted was increasing autonomy, coupled with increased federal funding. Indeed, the science policy of the 1950s and after had changed the nature and goals of scientists who constructed their enterprises in a world unknown to their predecessors. Federal funding and the centrality of the cold war with its massively expensive weaponry had inaugurated an era of what the director of Oak Ridge National Laboratory had called "Big Science," scientific research dependent on expensive machines, such as linear particle accelerators and mass spectrometers, to provide data. Physicists were especially addicted. Their World War II labors had helped inaugurate the era of federal money and into the 1960s they received by far the lion's share. More than any other single scientific group, physicists persisted as creatures of federal generosity as they worked in space, nuclear energy and weaponry, and defense. To these men and women, the status quo was barely tolerable, a drag on the progress of science and a slap in the face of the group that had hitched its star to the president. From their perspective, Johnson's failure to reward physicists with expansive, new, visionary programs demonstrated that the president was either merely a politician playing politics with America's future or a dolt who did not recognize nor understand the sanctity of science. In either case, the future for science—by which they meant physics—seemed bleak and required action on the part of all scientists to correct this appalling turn of events. If the president refused to budge, then the American people would have to be mobilized.

To be sure, the lack of consideration for the ambitions of scientists played a central role in their disillusionment with Johnson. Indeed, by their own words, they had initially decided to enter the political arena because Goldwater's nuclear policy could have ended civilization as they knew it. But this chilling prospect receded with Johnson's election. Few if any scientists expressed dismay about how the recently reelected president treated the nuclear option. Yet despite achieving what they claimed was their ultimate, appropriate, and self-disinterested goal, they nonetheless turned on Johnson because he neglected to reward them with increasing status, money, and programs.

Demands for a political payoff for its support were contrary to the explanation of why and how science received generous government funding in the first place. Science and scientists were explicitly supposed to be apolitical, possessors of a method that was to assist the nation and ultimately humankind; science was beyond reproach and not beholden to any political party or office. It was an approximation of truth, and this apoliticalness was one of the reasons that members of both major parties initially favored and supported federal funding of science. Science was in the national interest; it furthered the national interest and its practitioners' vessels through which to achieve that goal.

A group demanding political patronage could not remain on the moral high ground. Their quid pro quo stance reduced them to the status of political operative, seeking patronage as party loyalist, or as a vested or special interest. In either case, then, science would have become a client group, seemingly more interested in serving itself than furthering the national good.

As early as 1964 and certainly by the end of 1965, more than a handful of academics and others had begun to raise questions about the relationship between scientists and the public weal. Calling them "the new Brahmins," "the scientific estate," "the new priesthood," or some other similar thing, these scholars portrayed scientists, especially those in government service or seeking government funding, as privileged, responsible to no one, willing to push their agenda at almost any cost. One scholar pointedly mused that to expect scientists to construct a sound science policy had about the same chance as having "bricklayers make the best housing policy." These bureaucratically entrenched scientists—many of whom had made their mark as science administrators rather than as scientists—possessed a technocratic vision, a vision in which scientific sentiment was the sole logical means upon which to operate. No other group need be consulted because no other group possessed the skills necessary to decide; everything ultimately was a matter of science. Suddenly scientists were viewed as arrogant and inbred, and for a compelling reason. Though the initial congressional act creating the NSF called for an oversight board comprised of scientists from different types of institutions across America, by the later Eisenhower administration (and continuing through the Kennedy and Johnson administrations) nearly three quarters of the science advisors hailed from either MIT or Harvard. They issued their pronouncements, as if they were not subject to debate. In virtually all cases, these members of the academe called for the political process to rein in these powerful, ultimately self-interested rogues and to make sure that their techniques went to the public good. Rather than more latitude and flexibility, academics called for accountability and control by the political sphere.[16]

This sociohumanist harangue targeted the scientific elite. Others began to worry publicly about science and scientists generally. The atomic bomb, one of the themes of the 1964 election, seemed the epitome of the devastation unchecked scientists could wreak. This had been a long-standing critique both within science and outside of science. But it was a 1962 revelation that really seemed to channel energies. Rachel Carson's best-selling *Silent Spring* proved attractive at least as much for its form as its substance. Shunning statistics for narrative and dispassionate analysis for punch and verve, Carson used poetic imagery and vignette to portray a world in which science had run amok. She outlined how the insecticide DDT emerged as a scientific response to the plague of ants, mosquitoes, and the like, applied liberally because of its effectiveness. Carson took its proponents to task not because it failed to work but because of its profound social and environmental costs. DDT entered the food chain and affected species up to and including humankind. For Carson, it was this bigger picture, broader thinking, and more dynamic framework that scientists lacked, and that others outside that community must provide. Carson made the case against DDT with words, not numbers, and many people, especially humanists, recognized in her writing their newly found ability and newly discovered obligation to provide this service to society; looking for relevance, they found it in her work. In effect, Carson resonated with them because her book brought them into the mix. It allowed them to surmount C. P. Snow's exposition of *Two Cultures* and the contention that science alone furthered civilization. Rather than productive, *Silent Spring* had shown science and scientists to be barren and sterile without the contributions, caveats, or concerns of humanists and the humanities. Scientists were mere technicians, dependent upon others more broad-minded and ultimately wiser to fashion their productions into something vital and livable.

Both the stereotype of narrow scientists interested only in technical questions without regard to the world around him or her and of a scientific intelligentsia expecting to direct society suggested that it was a mistake to lump all scientists into a single discrete community. From the early 1960s, tensions among scientists had emerged. Among the clearest ruptures were between young and old. Professional organizations of scientists—formerly the arbiter of professional disputes—began to fracture. This was especially true of the American Association for the Advancement of Science (AAAS), the largest single scientific organization in America, and the AMA, where public spats were commonplace as was persistent criticism of professional policies. The nature of professionalism had changed in part to reflect the now intense relationship between science and the federal government. Even before creation of Scientists, Engineers and Physicians for Johnson-Humphrey, the AAAS had changed the nature of their journal, *Science*, to concentrate on policy issues. This was ironic because as late as 1962, the new editor, Philip Abelson, promised that he would maintain the status quo. He pledged to eschew editorializing, maintaining that he could "best foster *Science* by employing his energies on technical content." His only planned change, he vowed, was to make *Science* the journal of scientific record in all the scientific disciplines.[17]

In just one year the situation had become reversed. In an absolute repudiation of policy stretching nearly eighty years, the organization disavowed in 1963 any attempt to have the journal represent the profession and at a very public meeting explicitly granted it an independent voice. Part of the explanation for the decision of *Science*'s staff and the AAAS to opt for a journal that did other than record the progress of the profession stemmed from a trend new in journalism, the new journalism, the journalism of advocacy. This new journalism was explicitly not representative. It sought controversy. Its goal was to stimulate change and its enemy was the status quo. A particular favorite of recent journalism school graduates, the new journalism injected the journalist into the story as participant and/or adjudicator. The new journalism wanted to become relevant, to make a difference, to change what had been but by means that never were. Daniel S. Greenberg became *Science*'s first news editor and he devoted attention almost exclusively to what went on in Washington. Greenberg was ably assisted by several others, including Swarthmore College graduate Elinor Langer, who would later work at the radical chic *Ramparts*. Her editing approach favored those energized by Carson, who found themselves obligated to provide the bigger picture, including positing underlying themes and assumptions. Their writings usually were published in a section titled "News and Comment" as *Science* artfully joined the two and obscured what had been two quite separate enterprises. Covering congressional hearings, interviewing congressmen, and writing opinion pieces, Greenberg and the others made federal happenings the center of *Science*'s scientific universe. And indeed, so it had become to many scientists, whose research programs depended directly on federal funds.

Scientific exasperation with Johnson in 1965 also needed to be considered from a different angle. Certainly Johnson had more utility for more scientists and more different kinds of scientists than all his predecessors. Johnson's Great Society, his War on Poverty, and his attack on privilege all depended squarely on science and brought scientists into government. Research, even basic research, figured prominently in every Great Society program. Determining how to measure either a social state or progress in a specific dimension as an indicator of the efficacy of or the need to set social policy constituted in the parlance of the day "mission-oriented basic research." These new scientific beneficiaries of the Great Society and the War on Poverty were rarely physical scientists or engineers, two groups who in relative terms had had government funding in the 1950s and early 1960s all to themselves. Still, the number of physical scientists and engineers in federal service did not suffer a precipitous drop. What Johnson did was add large numbers of life and social scientists to the government rolls. If one did little more than count numbers of scientists receiving federal funding, then, it would seem that scientists were indeed rewarded for their political support. Many more of them gained government funding.

Not only did Johnson increase the numbers of scientists receiving government moneys, he also distributed decision-making power to more scientists. His willingness to use task forces—he created more than sixty during his presidency—enabled him to go right to the experts and to circumvent Washington's entrenched bureaucracy. Persons received positions on task forces not because they were ingrained as long-standing federal advisors but for other considerations, not the least of which were geographic diversity and reputation within the field. Johnson broke the Cambridge monopoly on presidential science advisors in another way. He replaced the Cambridge crowd with people working all across the country. Only one of his science advisory staff in 1965 hailed from that Massachusetts city. He appointed the first African American to the Atomic Energy Commission. Legislation Johnson sponsored also destroyed the inequity in the distribution of research funding. Under Eisenhower and Kennedy, twenty institutions, almost all of which were on the East or West Coast, received more than half the federal funding. Virtually every measure proposed by Johnson included a stipulation that funding be granted on a less-restricted basis, that "centers of excellence" be located throughout America.

These moves did not sit well with many scientists. Theirs was a professional mythology that claimed to distribute authority simply on merit, no other criterion. Power resided on the coasts and in Cambridge because persons there "did" the best science: they were smarter, more innovative, more productive. It was most efficient—scientific—to reward your best and brightest and thus reduce waste. Professional organizations and periodicals through offices, reputations, publications, and awards objectively selected the most meritorious scientists, and universities confirmed those decisions in their hiring policies. Under Johnson, a social vision of broad-based geographic and other kinds of equality supplanted supposedly impartial professional judgment. Scientists claimed that this social vision was ultimately political in that it served social purposes, not scientific goals.

That position was especially prevalent among those scientists who had been best connected and therefore had the most to lose. But Johnson's social vision also cut professional culture in another way. With his War on Poverty and Great Society, Johnson had elevated the social and life sciences to a status not unlike that reserved for the physical sciences since World War II. This was not simply about money; it was about self-identity and perception. To treat disciplines like sociology, political science, or other soft sciences similarly to physics was to physicists and many others perverse, a bold affront, injurious to most everything physical scientists believed. The result was not a campaign of introspection among the denizens of the hard sciences but rather an explanation for Johnson's actions that did not undermine the essential assumption of the physical sciences' superiority. To these men, Johnson either did not care or did not understand science. He was anti-intellectual, unfit for the presidency.

Not all scientists rejected the president. Many younger ones, especially social scientists, treated Washington as the center of the universe, but not just because of the money available there. Under Kennedy and Johnson, the federal government had adopted social policies that targeted privilege, aiming instead to provide similar high-quality services throughout the American population. This approach did not level the playing field but raised the bar for everyone except the most well-to-do. It depended on the skills of scientists but it also pointed to the critical importance of the political process. Young and headstrong and enamored with the quest for social justice, younger scientists sometimes found intellectual kinship with the Freedom Riders touring Mississippi and saw political mobilization, action, and change as the means to bring the virtues of science to the people. To these socially committed, politically attuned youngsters, Johnson's program was but a necessary first step in what would become the enduring federal assault on privilege and inequity.

These idealists were a minority. Many more scientists harkened back to the past where science seemed king. Again, physics and the physical sciences and engineering led the lament. Post-*Sputnik* policies had accelerated the production of these and other scientists. But Johnson's policies had produced no equally explosive number of new opportunities for federal employment and funding for them. For the first time in more than a decade, physicists graduating in 1965 did not have several jobs to choose from. In fact, not every recently minted physicist found employment immediately. Some needed to wait a year or so to secure employment.

The relatively modest employment constraints shocked and squeezed these scientists. Long accustomed to being in great demand, they confronted for the first time a situation that seemed uncertain. Ironically the matter was not as desperate as they contended. Continuing a trend that had begun during the Korean War and continued through 1967, Johnson, like other presidents, increased the federal science budget, including that of NSF. More money was available for scientists to frame their own research agendas. But the trend was less optimistic. While the budget had increased an average of nearly 15 percent annually since 1953, it now increased only about one-third as fast. Also, a larger NSF budget did not necessarily mean that there was more research funding for physicists. A greater proportion of NSF money went to life and social scientists than ever before; physical scientists now had real competition for federal dollars. And some funding was targeted, as in the case of the "centers of excellence program," which required nationwide distribution. Some money too went to programs outlined in the 1958 NDEA Act and reified in the 1965 Higher Education Act. Such was the case with a grant that went to two young chemists at the University of Wisconsin, who sought to revamp chemical education for the most promising students to create the next generation of Nobel Prize winners. Their grant enabled them to attract graduating high schoolers with outstanding test scores in chemistry and to design a new kind of chemistry education. Focusing on laboratory experiments and attempting to cultivate and teach creativity, the chemists separated their prodigies from other chemistry majors, established a separate curriculum, and assigned graduate students to mentor them. In this way, the recent high schoolers became sort of cadet chemists, seeing firsthand what chemists did and thought and how they did and why they thought it.

THE WAR AT HOME AND ABROAD

Increasing competition for federal dollars led to other forms of action. Some scientists tried to take their case for additional science moneys directly to the people but with little success. Science seemed to lack the dramatic cachet, the imperative it once had. As one gentleman put it, science might be "mighty, but not the almighty." Increasingly, people argued that science was but one claimant for public attention and one rival for the public purse. Warring against poverty and resolving the problem of the inner city—the neighborhood of Watts in Los Angeles exploded in August 1965, a riot lasting six days, leaving thirty-four dead, more than one thousand injured, hundreds of buildings destroyed—seemed much more immediate and much more important to the creation of a truly great society.[18]

So too did Vietnam. In August 1964 Johnson reported to Congress that a North Vietnamese ship fired on an American vessel. He asked Congress to authorize a response and permission to take all necessary measures to repel any armed attack against the forces of the United States and to prevent further aggression. By December 1965 more than 200,000 US troops were in Vietnam.

To the vast majority of Americans in 1965, Vietnam seemed an essential facet of the cold war. By meeting the communist threat in Vietnam, a greater, more dangerous, possibly nuclear confrontation would be avoided. At the heart of this sentiment lay something known as the domino theory. It supposed that communism was monolithic and that the free world was threatened whenever a nation fell under that system. In the case of Southeast Asia, the fall of South Vietnam would quickly be followed by a communist takeover of Laos, Cambodia, and perhaps Korea and Japan. Each falling domino would displace another domino until such time as the United States was directly and immediately threatened. To many cold warriors, the mantra "Better Dead than Red" was only too real.

Mobilizing forces cost money but so too did providing the military with the equipment necessary to undertake its tasks. Vietnam was a jungle war; aerial reconnaissance, quick, small mobile vehicles, helicopters, and the like needed to be designed specially for the terrain. In the instance of aerial reconnaissance, for example, a conventional aerial photo bath could provide real-time pictures of sufficient resolution to spot the enemy. But with few airfields large enough in the jungle to land planes of any size, airplanes needed to remain aloft for long periods of time, resulting in long photo reconnaissance missions and tens of thousands of feet of film. No one knew how the chemistry of the developer changed as all that film went through it. Chemicals in the bath were used in developing the film differentially. What was necessary to restore the bath? Was the restoration process linear or did it change at some other constant or inconsistent rate? Film chemicals—silver and others—poured off what became the negative and into the bath. This sludge changed the bath's chemical composition. How often did the bath need changing?—a question whose answer may have made aerial reconnaissance difficult, because it would likely require the aircraft to land. Did additions to the bath and sludge products interact, change

Fig. 5

American military airplanes spraying defoliants over Vietnam's countryside in an
effort to destroy the enemy's food sources
and prevent their troops from using dense
forests as ground cover. The US military
used an estimated twelve to nineteen million gallons of the herbicide mixture called
Agent Orange in South Vietnam between
the early 1960s and 1971. Agent Orange

was contaminated with highly toxic dioxin, and its overuse and abuse in Vietnam
poisoned water supplies and exposed both American troops and Vietnamese civil-
ians, a problem that some experts linked to unusual long-term health problems and
birth defects. Science and technology played a central role in many dimensions of
America's Vietnam War campaign, from the use of statistics and social science in
planning to the use of computers and new forms of weaponry.

---

meaningfully the composition of the bath, and affect photo quality?
These and numerous other questions were extremely complicated
and required extensive and sophisticated scientific investigation.
They also were critical to pursuing the war successfully. (See fig. 5.)

A proportion of science funding went to develop new techniques
and weapons to fight the war. Some of that money that had previously been earmarked for particular university campuses long
engaged in defense-related research never made it as several colleges
and universities debated whether its scientists should receive
defense- or military-related funding. This was especially ironic since
the cold war initially inspired federal science funding; without a perceived military threat, there would likely have been no government-
sponsored research money. Even though a fair share of universities
considered whether they wished to accept military research funding,
it did not signify a lack of support for the war. Very few Americans
opposed the Vietnam conflict in 1965. Yet the preponderance of
those who did were on college campuses. To these men and women,
Vietnam was "the wrong war at the wrong place at the wrong time."
Arguing that Vietnam was "an immoral war" and perhaps because of
their role in the creation of nuclear weapons (many of the original

participants were still alive), physical scientists were among the earliest, most active war opponents. Some started to march and teach against the war.

The simmering situation on campuses and the scientists' involvement in it did not go unnoticed in the White House. The more scientists led these still small antiwar rallies, the less clout Hornig had within the administration. By the end of 1965, Hornig had lost so much favor that the president no longer met with him on a regular basis. His inability to show scientists the virtues of Johnson's programs and therefore to contain demonstrations against the war made him seem ineffectual within government. Conversely, his strident defense of basic research made him seem to the very political White House simply a special interest operating within the administration. Rather than teaching scientists to follow Johnson's agenda, he was lobbying for theirs within the Oval Office at the expense of the administration's goals.

Scientists could no longer go to the White House with their grievances. And the war was another thing that pitted scientists against one another. It was not the sort of thing that a professional organization could adjudicate; it was not about science. With federal research budgets rising less steeply and employment opportunities less attractive, the situation seemed on the verge of spiraling out of control. The affront of not being spared from the political process where coalitions, ideologies, and partisanship ruled struck them as shortsighted and intolerable. Yet the only avenue to them was to make their case directly to Congress.

Congress, fresh from displaying their independence from the executive branch, greeted the scientists skeptically, as supplicants. One congressman acknowledged that "scientists and Congressmen have become interdependent," but he also reminded scientists that the growth in federal science funding had now given the "federal government . . . predominant control over the innovative capacity" of our nation. Such an important task required something more than ad hoc action. Congress energetically investigated establishing a nationwide science policy. Congress understood its members lacked formal scientific training and presently relied on staffers versed in the field.

Lack of a formal, permanent scientific staff meant Congress always was at a disadvantage to the White House with its full complement of science advisors. To at least provide Congress with its own dependable, independent source of scientific information, it created a Science Policy Research Division in the Library of Congress's Legislative Reference Section. This division did not set policy but gave Congress the information from which to establish policy.[19]

The broad outlines of this new congressional science policy became clear quickly. The product of an interdisciplinary effort, it would tackle societal problems, such as pollution and infrastructure, as well as other areas, such as nuclear nonproliferation. Funding would be distributed relatively equitably geographically. This all-encompassing policy would embrace the social sciences and ensure that they received due credit. Applied research and engineering would be priorities, as investigations not yielding solutions to America's myriad problems ignored national needs. Since funding was competitive, any policy must include some kind of measure as to the quality of research produced. Only in this way could funding fit national abilities with national necessities.

Representative Emilio Q. Daddario (D-New York) held a series of hearings on just what exactly this new policy unit would entail. As Congress deliberated, the National Academy of Sciences, the most prestigious scientific group in America, put together a fifteen-member panel to bring scientific opinion to the science policy question. What resulted, however, were fifteen separate opinions. While the committee concluded that the NSF should continue to fund "basic research—especially in the physical sciences—that is too remote to merit support from the mission-oriented agencies," its members failed to agree on how to frame policy, who should frame it, and what it might entail. In effect, the committee demonstrated the critical tensions and fissures within what had seemed a decade earlier a harmonious, relatively homogeneous scientific community.[20]

These cleavages were not easily sutured together. Congressional hearings exacerbated the matter. For the first time, prominent scientists considered the reality that scientific funding was limited and that choices had to be made among various worthy projects. These men decided publicly to offer Congress advice rather than leave the matter to chance. High-energy physics was among the sciences taken to task. While Hornig supported the field and claimed that a new huge linear accelerator was necessary for American preeminence, others disagreed. Money would be spent there with dubious prospects while more immediate projects would be denied. Abelson made the case succinctly. "Never, in the history of science have so many fine minds been supported on such a grand scale . . . and returned so little to society for its patronage." The mistake was obvious. Persons who recommended spending money in this fashion were high-energy physicists. That was "like asking a hungry cat to make recommendations about the disposition of some cream."[21]

Congress did not establish a science policy entity in 1965. But the Daddario committee did recommend changes in the NSF's charter and the way it operated. To the committee, the NSF must take the policy lead. It chided the organization for failing to follow its authorizing act empowering it to develop a national policy for research and to evaluate research being undertaken within the federal bureaucracy. It demanded that the NSF support applied research, distribute funding on a geographic basis, and incorporate social science research more neatly within its activities.

Despite Congress's inability to frame a science policy alternative to the presidential advisor system or even revamp the NSF, its authority over federal science actually increased in the later years of the Johnson administration. Hornig remained on the outs with Johnson as anti-Vietnam demonstrations on college campuses rose to a fevered pitch. In 1966 and after, Johnson rarely mentioned science in his messages to Congress. When he did speak of science publicly, it was rarely complementary. Perhaps his most-publicized initiative was his medical research summit. On June 26, 1966, he called together the directors of various federal medical research agencies to meet at the NIH. There he reminded them that the federal government had spent billions sponsoring medical research, and he demanded to know what there was to show for it. "The time has now come to zero in on the targets," he warned ominously. The implications were clear. Research needed to pay off now.

Johnson's emphasis on immediacy and limits resonated in late-1960s America. Inner cities were poised to explode and did during 1967's Long Hot Summer. Regulatory agencies seemed unable or unwilling to protect Americans. Concern about DDT was quickly followed by Thalidomide, a drug that produced an epidemic of flipper-limbed babies. A young Ralph Nader first exposed the automobile industry in 1965's *Unsafe at Any Speed* and then meat processing the next year. Capitalizing on his popularity and the fact that others shared his concern over the failure of the government to protect the citizenry, he gathered an idealistic cadre of young men and women to investigate governmental regulations generally. In 1969 they became known as "Nader's Raiders." A year earlier, a task force headed by Hornig on the quality of the environment noted a marked decrease in environmental conditions and a deteriorating situation. Younger scientists and doctors became increasingly disappointed with their professional organizations. A smaller percentage of recently minted scientists joined professional organizations and many who did found them outmoded. A similar phenomenon occurred outside of the scientific community as the split between young and old Americans was becoming clearer and wider. In 1966 *Time* magazine recognized this expanding chasm by naming "the Younger Generation"—"not just a new generation but a new kind of generation"—its "Man of the Year." The ever-escalating Vietnam situation drew more resources and engaged more university students in increasingly strident protests. Despite these disturbances, more than 540,000 American soldiers were stationed in Vietnam during August 1968. Society seemed to be coming apart at its seams. Visions and promises gave way to hopes for stability and disillusionment.[22]

Johnson labored to fund Vietnam and to maintain support for Great Society programs and the War on Poverty. Some, like the regional medical centers, were cut drastically and virtually ceased to exist. Yet the White House managed to propose a new program, Project Sentinel, to erect antiballistic missile systems around major American cities just in case Mutually Assured Destruction would fail to dissuade the Soviets from a first nuclear strike. Congress refused to grant the administration the final go-ahead for Sentinel. That body too was imbued with a sense of limits and instead it endeavored to streamline and systematize federal activities. Congress wrote rules by which universities kept a tighter rein on NSF and NIH grants, which required them to certify that funding only went to the projects to which grants had been awarded. Congress also began to take issue with some of the projects sponsored. Social science funding was especially scrutinized. Pointing to "the federal research craze," congressmen termed it a "profitable parasite industry" and maintained that many social scientists were "getting fat at the public trough." To better decide what programs should be funded, some congressmen called for creation of a technology assessment unit to provide cost-benefit analyses for projects seeking federal funding. Such a move was necessary, according to a congressional report, because of the "dangerous side effects which applied technology is creating, or is likely to create," effects "so strong—and quite possibly so dangerous—as to pose a genuine threat to man and his physical, mental, and spiritual environment." Technology assessment, of course, was antithetical to basic research, from which no one could know the possible spin-offs or applications. So too was another venture, this time under the banner of the AAAS. It held a large symposium to discuss "How Man Has Changed His Planet." Speakers noted changes in weather, water and air quality, creation of pervasive chemical pollution as a by-product of industrial agriculture, and how Europeans viewed nonwhites as well as how "Christian arrogance toward nature" produced many of today's problems. The famous geneticist Theodosius Dobzhansky summed it up best when he emphatically said that "in our world, a scientist has no right to be irresponsible." The audience thundered applause.[23]

How humans changed the earth depended not on planning but on what had been attempted and accomplished. Yet this change seemed to highlight the need to plan, to fashion some sort of unified science policy to prevent other unanticipated breakdowns. To that end, Congress began a campaign to coordinate and simplify the interlocking and sometimes redundant committees and subcommittees charged with overseeing the scientific enterprise. Yet this simple chore was a dismal failure. In an organization where huge amounts of power rested in reporting measures to committees and committees reporting to the floor, no congressman volunteered to cede their authority for the good of the whole.

It was not surprising, then, that scientists refused to take the federal funding slowdown with good cheer. As Greenberg noted in *Science*, a tendency existed among "some scientists to equate the refusal of a grant with the persecution of Galileo." Others saw that "science is going the way of Studebaker," a recently bankrupt automobile firm. This came from the fact that in the past two decades, federal programs had "created a vast population of consumers of research support, all imbued with the spirit of rising expectations," Greenberg acknowledged. The result was despair among some, "while others are beginning to experience the scale of professional living normally associated with the English department."[24]

In 1968 the worst happened. For the first time since the Korean War, the federal science budget did not rise. As important, it was cut differentially. Both the NIH and the Atomic Energy Commission got less than expected but held their own. The NSF received a modest increase, but with the stipulation that the agency focus on social science, chemistry, and atmospheric and oceanic sciences. The NASA budget was reduced by about 10 percent. Basic research in the Defense Department, even "mission-oriented basic research," was slashed by about 12 percent. Within defense, the area hardest hit was the Advanced Research Project Agency. The "striking reduction" of the ARPA budget almost terminated a project begun in 1962. The agency's new director, MIT's J. C. R. Licklider was given the task of integrating a new device, the computer, into the military. Over the next six years, he proposed creating a network of computers over a distance and worked with his staff to develop standardized protocols and methods for transferring data. Finding that the research could be speeded through contracts with universities, Licklider negotiated agreements with several western universities to form what was then known as the "interface message processor."[25]

Contracts signed before the new budget year remained in force and Licklider's network survived to become the forerunner of the modern Internet. But it was an exception. Regional medical programs and environmental health services, two Great Society programs, were held in abeyance as Congress argued that neither was established enough to consider expansion. Despite Congress's unwillingness to continuing funding these ventures, both the environment and medicine had done well during Johnson's second term. During that period, Congress approved nearly three hundred environmental measures and authorized $12 billion in funding. Under Johnson, federal funding had helped more than 100,000 doctors, nurses, and dentists receive training, an additional 123,000 hospital beds had been opened, and a dramatic 13 percent decrease in infant mortality had been achieved. Nearly half a million poor handicapped children received treatment that they otherwise would have gone without.

Physical scientists felt they could make no such case. Their area took a significant budget hit. More important, it seemed as if the situation they faced was permanent. The dramatic rechartering of the NSF in July 1968 undermined their intellectual position further. Reworked by both congressional houses and the White House, the NSF was for the first time subject to annual reviews of its programs by the House and the Senate. The agency must convince Congress of the efficacy of its activities if it was to receive funding for the next fiscal year because the recharter now required the NSF to seek an annual appropriation rather than depend on a continuing one. The White House gained accountability for the organization's executive functions as the president—not the NSF director—now chose the organization's four assistant directors. The recharter specifically designated social science as eligible for federal funds and explicitly approved the idea of supporting applied as well as basic research, two measures that potentially diminished even further possible moneys available for research in certain branches of the physical sciences.

Redrafting of the NSF's charter was the natural intellectual consequence of the thrust of Johnson's administration and Congress's participation. The Great Society and the War on Poverty sought to apply scientific techniques to resolve social issues. They were not about uncovering new principles to make new weapons, discovering new devices to show the material superiority over the Soviets, nor designing new media upon which to send messages. They were about curing disease, applying social science methods to real-world tasks, and producing a hardier seed corn. In this world, the application of science was more critical than the production of science and overwhelmed or overshadowed it. Techniques, mechanisms, programs, and machines, each designed to come to bear on a specific issue or problems and only designed with that in mind, became a crucial facet of federal science funding. Technology had emerged within the federal bureaucracy as more compelling than science; immediate applicability seemed more pressing than possible future breakthroughs.

This manifestation of a sense of limits revealed the future as more distant and less certain, and therefore less urgent. Emphasizing immediacy, coupled with a potent sense of limits, converted measurements of the effectiveness of a technology into an important social goal. Technology was today, relevant. Science was the future, tomorrow, which in the parlance of the time, may be canceled due to lack of interest. Tune in and turn on to technology.

THE ELECTION OF 1968

Johnson's decision not to seek reelection in 1968 provided an opportunity for the reputed American scientific community to mobilize its support behind a single candidate. There would be no shortage of political action in 1968. Thousands of scientists participated not as individuals but as scientists in the election. But their collective influence was less than it might have been. The massive fissures in American science and society prohibited a single cohesive strategy or even the backing of a single candidate.

Johnson's Vietnam policy disturbed younger scientists, who lent their support to Eugene McCarthy. Running on an antiwar platform, the Democratic senator from Minnesota pledged a swift and sudden withdrawal from that Asian country. McCarthy also attracted American science's old guard, many of those scientists still atoning for their involvement in the Manhattan Project. Backing McCarthy was extremely high principled. Although many no doubt felt comfortable with his academic demeanor, his record in the Senate and else-where with respect to science was virtually nonexistent; only faith suggested his support of science funding would not be dismal.

Hubert Humphrey, who would ultimately beat out McCarthy for the Democratic nomination, had a stellar science record. But he refused to renounce the policies of which he was a part of the pre-vious four years. Few McCarthy supporters would in good conscience offer public backing to a candidate approving of the Vietnam con-flict. Also, Humphrey seemed too old-fashioned for the younger gen-eration. As part of the Washington establishment, he was considered by some to be part of the problem, not the solution. When persons talked about the system not working anymore, they were referring to the relatively slow pace at which government moved to mitigate social problems and the tired methods it employed. The debacle at the Democratic National Convention in Chicago cemented Humphrey's establishment reputation. There, Mayor Richard Daley unleashed the Chicago police to quell demonstrations undertaken to protest the war and to disrupt the convention. The riot that ensued was televised nightly for much of the week and fixed Humphrey as a card-carrying member of the establishment.

But Humphrey did draw to his side those scientific politicos, men like Donald MacArthur and Jerome Wiesner, who had long been involved in federal science policy formation and implementation. Together this group worked to rekindle the spirit of four years earlier and established a Scientists and Engineers for Humphrey-Muskie group. This group was rife with problems from the beginning. It managed to list several prominent men—Nobel Prize winners—who had not offered Humphrey support. Each denial and repudiation undercut the group. So too did its members. McCarthy's scientists had outnumbered Humphrey's Nobel laureates twelve to one. Most of Humphrey's supporters were in the social and life sciences or were doctors, those groups that had done well under Johnson.

The MacArthur-Wiesner revival never gained much momentum, in part because several prominent scientists recognized that political partisanship was self-defeating. "Organization of partisan groups of scientists supporting individual candidates for high political office threatens to generate serious rifts in the scientific community," noted Philip Handler, then head of the oversight board for the NSF. It also would undermine peer-review conducted science funding because it would be "inevitable that national attitudes and federal support for science must also come to involve political considerations." This was "ill-advised and a disservice to our society." As a result "our nation will suffer."[26]

But it was difficult to put the genie entirely back into the bottle. Experiences during the past presidential election carried over to 1968. Not to be outdone or caught unaware, the Republican nominee, Nixon, created a Committee of Scientists and Engineers to demonstrate that members of both communities supported his candidacy. Comprised of military men and industrial and academic research directors, this corporate group specialized in nuclear energy and weaponry as well as commercial development of science. Created less as a political organizing body than to advise the candidate, the group quickly proved anathema to scientists. Not only were few of these men noted for their scientific discoveries or breakthroughs, but they included prominently several former Goldwater supporters. Backing Goldwater in 1964 had seemed to many scientists irresponsible. In the case of Nixon, his political activities overshadowed his science policies. Scientists as a group had a visceral dislike for Nixon. As part of the House Un-American Activities Committee and having seized an active role in the Alger Hiss investigation, Nixon seemed antithetical to the spirit of democracy and of science. To many scientists, Nixon as the most successful and enduring politician of the era symbolized McCarthyism. He was to them the repository of all things they despised from that period: scurrilous investigations into scientists and others, paranoid delusions about lack of loyalty, tarring reputations permanently with innuendo, restrictions on academic and other freedoms, and blacklisting because of supposed political beliefs or affiliations. In this scheme, the denial of a security clearance to Robert Oppenheimer in the mid-1950s reigned supreme. The head of the Manhattan Project, Oppenheimer's second thoughts about the wisdom and morality of nuclear weaponry made him a target of suspicion. Denying this hero of scientists a security clearance was the ultimate affront to much of the American scientific community. That Nixon had no direct part in the refusal mattered far less than his active participation in the era in which the rejection was so much a part.

Ironically, neither Nixon nor Humphrey mentioned science explicitly during their nomination addresses. Both accentuated the need for law and order, but their methods differed. Humphrey would achieve the goal by extending the programs initiated by Johnson to increase opportunity and to diffuse the blessings of America through a wider swath of American society. Nixon wanted to reestablish the American Revolution where "the first requisite for progress [was] order." Reestablishing domestic tranquility would be followed by unburdening middle America—the law-abiding "silent majority"—from federal interference and thereby enabling it to increase prosperity generally through America. The spirit of individual initiative would sweep the land and provide solutions to the vast majority of social problems.[27]

Both candidates were more candid on the stump. Nixon even issued a full-scale address titled "The Research Gap: Crisis in American Science and Technology" to lay out his case. Delivered on October 5, the detailed and specific address claimed that the American scientific community was being "shortchanged," that a "research gap" was emerging between the Soviets and America, and that the current administration hobbled science by asserting that we have reached a "technological plateau," a situation where "technological potentialities are limited." To counteract the gap, Nixon promised increases in basic research as part of a "new national commitment" to science. "Scientific activity cannot be turned off and on like a faucet," he warned, and he mentioned that new military weaponry required extensive research before development. While Nixon backed a nuclear test ban, he also contended that America needed to achieve military superiority, not equality, with the Soviet Union. To aim toward parity left too much to chance and no room for error. Misconceptions and unanticipated circumstance could derail America's armaments and therefore enable the Soviet Union to establish preeminence. Nixon also championed breeder reactor research as an idea whose time had come. America's dependence on foreign fuels would leave it vulnerable to blackmail or worse. Nixon criticized White House science advisors for wasting their efforts attempting to construct a grand science policy, arguing that those efforts were secondary to the more mundane duties of managing, implementing, and otherwise carrying out policies already in place. He opposed a federal "science czar" for that very reason. Nixon pledged to continue the moon effort but pointedly did not promise new major space initiatives. Finally, he favored using federal science to measure social and economic problems and then enlisting private enterprise to help resolve those issues. Federal social science could then be employed to determine the efficacy of those remedies. Finally, Nixon argued for new research into the treatment of the mentally ill as well as "lasers, pollution research, and computer technology."[28]

Humphrey relied on his record rather than outline a science plan until nearly a month later. Titled "Science to Serve a Nation," Humphrey's plan emphasized how science, especially the "behavioral sciences as well as the natural sciences and engineering," could be brought to bear on the issues of "crime control, slums, pollution, housing, health care, hunger, and education." In addition to focusing on what "science can do for people," a direct homage to Johnson's Great Society, Humphrey supported nuclear arms control and saw parity as the first step in achieving it. Parity meant that both sides could negotiate from strength; neither felt compelled to catch up or to compensate for an inferior position. He also expressed a desire to create a cabinet-level department of science and technology. In this unit, all the various scientific and technical activities of the federal government would be housed, thus providing synergy throughout government. As significant, its cabinet secretary head would be directly involved in all presidential decision making. Humphrey wanted to reduce costs of spaceflights but maintained the need to continue research in that sphere. He concluded by calling for expanded oceanic exploration and "multi-disciplinary technological institutes on urban sciences, transportation and environmental management."[29]

Both Humphrey and Nixon saw science as something applied. Humphrey accentuated social issues and Nixon highlighted the military. To be sure, Johnson had also moved to apply science. But with both Nixon and Humphrey there existed considerable differences from Johnson. Both candidates in 1968 were less interested in the future direction of scientific efforts than in their responsibility to manage scientific matters better and to apply them more quickly and consistently. Efficiency mattered as much as efficacy and more than conquering new frontiers or erecting a greater society. At a time where people were proclaiming that "tomorrow had been canceled due to lack of interest" and that "the future is now," immediacy took precedence and application bested investigation. To be sure, a few scientists long active in partisan politics continued to battle about basic research—Wiesner, for example, called Nixon's science team representatives of "the troglodyte, or dinosaur wing of the scientific community," and argued that anyone seeking their advice "is not a man you'd want to be president of your country"—but their numbers were greatly diminished as was their influence. It was the beginning of the end for the World War II generation.[30]

WHITHER SCIENTISTS

Science greeted Nixon's election as business as usual as it speculated about who would fill the openings in the federal science bureaucracy. But increasingly the status quo seemed an untenable proposition. Resisting political action would continue what seemed to many scientists as the decline of science's influence. Daddario urged scientists to involve themselves directly in the political sphere. Arguing that continued science funding depended on the establishment of political coalitions favoring it, that modern life was imperiled by unforeseen consequences of new technologies, and that only new knowledge could overcome the morass of contemporary existence, he called on scientists to become an integral force in US politics. In this way, scientists could introduce "reason as a balance for emotion" and "solve the social problems which are the ultimate source of the unrest." Don K. Price, a Harvard political scientist, made the case simply. Noting that the Romanticism of the previous two centuries neared its end and America would be left with a series of hard choices, he demanded that scientists join the public discussion "to help clarify our public values, define our policy options, and assist responsible political leaders in the guidance and control of the powerful forces which have been let loose on this troubled planet."[31]

The nobility of science resolving current-day problems was matched by an equally pervasive rhetoric that science and by extension scientists were responsible for contemporary problems. Philosophers and other humanists, such as André Malraux, Jacques Ellul, Erich Fromm, and Herbert Marcuse, were read and taught at universities. To these authors, science was blameless but scientists were not. They had sold their souls for profit to capitalism, the modern state, multinationalism, or some such other overarching predatory system. More practically, rarely a day seemed to go by without another revelation about some atmospheric pollutant, carcinogenic substance, or chemical in drinking water.

These sorts of analyses rendered it less important that a person was a scientist than to whom the scientist owed allegiance, a potent acknowledgement that scientists did not constitute a cohesive community, but rather various interest groups. American science as a singular, unified, coherent entity was a sham. Within that context of an American science with numerous, conflicting allegiances and affiliations, who guarded, who guaranteed, who spoke for the public interest? Who ensured that it be served?

Academic institutions, the recipients of much of the federal largesse, recognized that continued federal money depended on political action. National Association of State Universities and Land-Grant Colleges, the lobbying arm of large public schools, took the initiative by drafting a bill that would approve supplemental institutional grants—grants directly to colleges and universities for support of science done there—in addition to grants to individual investigators for specific projects. Pegging this additional support, at least 20 percent of the federal research budget, the universities sought to become a decision-making body for federal funds; granting money to institutions would supplant peer review within those institutions. It also would tie investigators to the institution, as the funding rested there, not with the investigator. Funding for this venture would be handled by the Office of Education, not the NSF, and distributed primary on volume of federal money already received and the number of advanced degrees given out.

Institutions that would not fare well under this rich-get-richer formula objected but generally supported the idea of individual institutional control of federal research support. This measure did not gain broad congressional backing, however, in part because decentralized control of research funding seemed antithetical to a coordinated national policy and because the measure provided no means to guarantee accountability. A more common assessment was that scientists were already generously funded and to expect more during tough budget times seemed greedy, irresponsible. It was with that sense of dissatisfaction that Senate Majority Leader Mike Mansfield's (D-Montana) proposed measure to limit all defense appropriations to those things having direct and immediate military application gathered considerable congressional support. The Senate approved the measure but the House did not consent. If it had passed, a huge section of the federal science research budget would have been removed.

Ironically, as Congress debated Mansfield's measure—it would become law with the 1971 budget, reducing military science spending by an estimated $50 million—faculty at several prominent research universities considered the morality of accepting defense-related funding. Scientists and others at MIT, Stanford, Cornell, Columbia, Yale, Carnegie Mellon, Rutgers, Northwestern, Minnesota, Maryland, and California celebrated March 4, 1969, by considering the relationships among scientists and their productions. Most claimed it was the scientists' responsibility to ensure that governments did not misuse science, while others demanded that research concentrate on those issues to elevate the human condition. Some maintained that dressing bad policy in the shroud of science insulated it from public scrutiny and that scientists needed to uncloak those charlatans, while others maintained that "our government has become preoccupied with death" and contended that its policies were simply "criminal insanity." And they attacked the venerable federal science institutions—NAS, National Science Board, and Office of Science and Technology—as the "worst offenders," as the backbones of the military-industrial complex, as nothing more than merchants of death.[32]

These turbulent meetings did little more than clear the air. Stanford was an exception. A nine-day long sit-in led to the campus phasing out most classified research and establishing guidelines to bar almost all new military research. MIT also modified its procedures, though not as significantly as Stanford. It agreed to a moratorium on research in two of its major military research labs. Disruption and disaffection were not limited to universities. Major scientific and medical organizations lost as much as a third of their membership in the later 1960s and early 1970s as these organizations seemed increasingly irrelevant. Sometimes new organizations emerged, such as the Radical Physicians, who pledged to work at inner-city hospitals to improve the lot of the poor. Annual meetings sometimes assumed a confrontational character. At the Boston 1969 AAAS, for instance, a panel to discuss "The Sorry State of Science—A Student Critique," led to a generalized disruption of sessions, excoriation of scientists for supporting the arms race and the space program, and shouts of "Science for the people!"[33]

Even more common in 1969 was an orgy of analysis about what could be done to federal science in America. A new president was about to take office. Few public figures expressed satisfaction with the status quo. All offered their ideas of ways to improve what they recognized as an outmoded, ineffective system. These analyses occurred on many levels but were pervasive within government. The NSF, for example, established a model for determining the relationship between basic research and breakthrough technological innovation. This time series demonstrated to its already convinced proponents that the period two to three decades prior to the innovation was most critical and that 70 percent of the key events in technological innovation were the direct result of nonmission research. It concluded therefore that "undirected research, with knowledge as the only goal, provides a reservoir of understanding essential to subsequent technological innovation." The Department of Health, Education, and Welfare (HEW) called for a "comprehensive set of statistics reflecting social progress" and "social indicators" to gauge the effectiveness of the government social science efforts. Arguing that government has simple measures of life, death, and achievement, HEW now wanted to measure happiness and satisfaction. Only in that way could government learn about the "efficacy of public programs" and make "informed judgments about national priorities." The public has "no measures of the satisfaction that income brings." It has "no way to tell" whether the environment is "becoming uglier or more beautiful." It has "no measures of physical vigor." Rather than record life expectancy, HEW argued that it was more meaningful to consider instead "the expectancy of healthy life (years free of disability requiring institutional or permanent bed care)." Similarly, it seemed less useful to talk of criminals rather than "the extent of criminality." Only through "the reduction of social well-being to statistical form," only through "what ought to be known," could America make effective public social policy. Hornig, who was leaving the Office of Science and Technology, reported that the office should be strengthened and that while he realized "that science is part of everything," nonetheless favored a new cabinet secretary of science. Cabinet status would grant the secretary "line responsibility and public accountability and, most importantly, the interest and confidence of the president, the attention of the Bureau of the Budget, and the ear of the Congress." It would become an "office of planning, evaluation, and analysis, looking broadly at national problems with some scientific or technological component but extending well beyond the purely technical areas." Systems analysis—rational analysis and experimentation—would replace "bureaucratic and political processes" so the virtues of American science would be felt throughout society. James A. Shannon, head of the NIH, argued that in addition to a more regularized public relations campaign, the NIH must develop "central analysis and planning functions that are adequate to the task of ordering national priorities and serve as a basis for the allocation of resources" within the biomedical fields.[34]

Each of the aforementioned self-serving statements demanded the continuation or expansion of programs and processes already extant, yet each offered justifications or techniques that would make agency activities seem new. Desire to dictate the future by relying on established parameters appears to suppose the status quo, but in these instances it did not. Rather than attempt to confront the unknown, a future more limited than the present, they offered more of the same. Each proposal understood that public opinion would become a potent factor in federal science and technology, and each provided a way to maneuver opinion. Public relations campaigns, overhauls with an eye toward efficiency, introduction of new metrics, and dependence on social science studies—science—of public action emerged as the bailiwick of national planning as well as a bulwark against the indeterminate future. This was as true outside the government as in it. Yet each analyst dressed the arguments they offered as science, as rational and objectified, as arrived at without any a priori assumptions. Resolutions they offered were therefore facts; no room existed for debate. Anyone who opposed a plan was unscientific and recommendations by such persons were not worthy of debate.

Science-dictated and -determined planning would enable the American people to plan the future to maximize the benefits of science. But that was a strategy that no longer made sense. As in the case of military or classified research, scientists often took contradictory positions and vehemently opposed one another on many issues. And when scientists disagreed—their authority stemmed from their ability to get reproducible results and therefore to offer dependable, consistent statements—anything that they might have had to say due to their technical skills was rendered irrelevant. Eugene Skolnikoff, an MIT political scientist, recognized the new way. Congress had developed its own science reference section and "citizens' groups . . . make independent analyses," while the OST "challenges agency positions on complex technological questions." Each of these groups represented its members' perspective. What was necessary was an independent science voice, which Skolnikoff thought had been the traditional role of the university. It had been "the primary locus in our society for critical examination of social issues from a base of strong analytical capability characterized by a striving for unbiased scholarship." Yet he also worried that the university's dependence on federal funding made its independence on many issues impossible. It had become the creature of the entity nurturing it, a special interest just like the rest.[35]

SCIENTISTS TO THE BARRICADES

Nixon's decision to revamp Johnson's antiballistic missile initiative sparked one of the first postelection outbursts. Project Sentinel became Project Safeguard as Nixon focused on defending America's nuclear weaponry and therefore its retaliatory potential rather than its cities. For this decision he relied on physical scientists within the military and the Defense Department who assured him it was technologically possible, and affiliated social scientists who argued that it enhanced American strength without exacerbating USSR–USA relations. To ensure that the Soviets did not perceive Nixon's new decision as an immediate threat, he stepped up the pace of the Strategic Arms Limitation Talks begun under Johnson, which aimed to reduce the possibility of a nuclear holocaust.

Almost immediately several former science advisors saw the reconfigured ABM strategy as a dismal mistake. They found a receptive audience in the Senate Foreign Relations committee, who felt that Johnson and now Nixon dictated to them and had eroded the body's right to give advice and provide consent. Through the forum of a public hearing, the advisors disagreed with the administration's experts and provided their own assessments. At the hearing's conclusion, a congressional aide offered this evaluation. "In opposing the ABM, the scientific community has come into its own as a political force." That act, he contended, "is science's most golden, glorious hour." The statement was more than hyperbole, of course. It was downright wrong. The scientific community had not opposed Nixon's ABM strategy. Individual scientists did. Individual scientists had also helped author Nixon's approach. There was no glory. There was no coming of age as a political force. What there was, however, was open co-option, the seizing of a group's mantle by an agenda-riddled congressional aide simply as support for a case that rested well outside the group's expertise.[36]

The World War II physical scientists predictably opposed the ABM. In fact, the issue, along with Vietnam, gave them an opportunity to res-urrect their moribund Federation of Atomic (now American) Scientists. With three thousand members just after the war, the organization had shriveled to about one thousand. But now it decided to hire a full-time lobbyist and to engage in a membership drive. Lobbying methods were straightforward. An older, publicly visible member—generally some-one who had worked on the Manhattan Project—would meet with a select group of congressmen for a breakfast/informational seminar in which the venerated figure would lecture on the flaws of the ABM con-cept and argue that no research gap existed between American scien-tists and their Soviet counterparts.

Other individuals, most commonly scientists with academic posts, joined the campaign to convince Congress to veto White House plans. A petition signed by 1,100 physicists was presented to the administration. But the matter was more complicated. At the annual meeting of the American Physical Society in early 1969, a group favoring rejection of Nixon's plan surveyed the group and found that more than one-fifth of their colleagues not only dis-agreed with them but also actively backed Safeguard. Would one group or the other represent scientific opinion? Who would decide? How would each group identify themselves?

The type of fissure exemplified by the ABM debacle exemplified science in the later 1960s and early 1970s. For scientists to have influence there needed to be a scientific viewpoint; scientists needed to agree since their method precluded the possibility of disagree-ment on scientific questions. But what would happen if two or more scientists examined a situation and came up with radically different understandings? More significant, policy is about the future, a realm outside of scientific abilities. Science, of course, can determine nothing about the future; it can only test hypotheses. So how would various hypotheses be tested for or about the future? What if one group of scientists argued that little evidence existed that a serious scenario might occur while another maintained that a particular act almost certainly would cause an extremely hazardous situation? Was that not the ABM situation? Who would judge? When disagreement surfaced, did that indicate that the matter under consideration could not be resolved by scientific means and therefore was not a scientific question? Was not the wiser course to stop the particular act—whether causal or not—because the consequences of one position far outweighed the other? If that were the case, then the reality of the situation did not matter. What mattered was the rhetoric and the implications offered by that rhetoric. Decision making was little more than marketing, advertising, and scientists—pitch men and women—shills for their emotional predilections.

There existed no scientific body to adjudicate what in essence were not scientific issues. The public was left to its own devices. It needed to factor its myriad politics, philosophies, interests, reputations, and a host of other imponderables. Nixon, in a sense, was the first president to operationalize fully this understanding about decision making and the role of expertise. His solution was simple. Only those persons with political beliefs and a world outlook similar to his needed to be included in presidential deliberations. One could find good Republican scientists as well as good Democratic ones, and Nixon embarked on this strategy by appointing people he knew and trusted to advisory panels in science, space, health, resources, and the like. He wanted what he thought were fresh perspectives—what he thought was a new, honest look devoid of the partisan—unabashedly liberal—considerations of the previous decade. What Nixon understood was that he himself possessed positions, political philosophies, assumptions about how government worked and human nature. What he lacked was technical facts. And to Nixon's administration facts were essential. They would demonstrate if something were or were not feasible, perhaps even necessary, and advisory committees did not recommend but instead provided a potpourri of options from which the president and his closest advisors would choose. Politics determined if it was desirable and therefore under what circumstances and conditions it should or should not be done.

This ought not suggest that Nixon held a personal animus

against science. He found himself regularly impressed by its power to manipulate the physical world, the essence of his position in the Kitchen Debate nearly a decade earlier. When the manned lunar landing came to fruition on July 20, 1969, Nixon shared the television screen with the astronauts for an uncomfortably long time and used the occasion to signal America's scientific and technological preeminence. But he also recognized science as a means, not an end, something that humans did to achieve a particular goal. Rather than possess an independent existence, Nixon saw science as a strategy for achieving something, one of perhaps many possibilities, and therefore in competition with those possibilities. What he had difficulty with was not science but scientists, particularly academic scientists. He refused to grant scientists "any special or privileged role in national life." Their sanctimoniousness and self-serving—Nixon thought it self-delusional—rhetoric about their needs and the government's obligation to provide them reversed what Nixon recognized as the proper relationship between scientists and government. Scientists operated for the national interest; it was not necessarily the national interest to cater to science.[37]

Scientists needed regular channeling; their often-unrealistic projections, complaints, and suggestions required constant scrutiny. More than any president before or since, Nixon vested his Office of Management and Budget with the task of modulating scientists; it was through that office that all options for domestic matters—scientific or otherwise—were vetted. Enough of a main street Republican to want to balance the budget or at least minimize the deficit—he instituted wage and price controls rather than reduce the size of the budget dramatically—Nixon subjected scientists' plans, proposals, demands, and the like to overwhelming inspection in what constituted a de facto cost-benefit analysis. In effect, those things high on Nixon's ideational/political agenda did better than those with which he had less of a commitment.

Nixon's science advisor, longtime friend and supporter Lee DuBridge, had had an impressive career as a research physicist at Cal Tech—he had been involved in creation of the NSF in 1950 and worked on its oversight board for years—and had served as president of that institution for more than two decades. But once in Washington, DuBridge encountered what seemed an almost impossible situation. His appointment signaled his ability as a scientific administrator and his loyalty to Nixon's agenda. But scientists had long treated the OST as the chief representative of their interest. DuBridge needed to satisfy scientists, badly fractured into numerous subunits, and the president.

Difficulties surfaced almost immediately. The National Science Board nominated Franklin A. Long, vice president for research at Cornell University, to head the NSF. A member of both Scientists, Engineers and Physicians for Johnson-Humphrey and Scientists and Engineers for Humphrey-Muskie, Long had long opposed any ABM system. DuBridge supported his nomination but Nixon vetoed it on the grounds that his opposition to ABMs could provide disaffected congressmen with a cudgel to bludgeon the president. Horton Guyford Stever, a physicist who headed Nixon's preinauguration task force on science and technology and president of Carnegie Mellon University, explained the president's position succinctly. "No administration," he complained, "can withstand within itself an activist against itself." Long's ABM testimony would embarrass the White House and become the focus of congressional dissent. The overwhelming majority of establishment science—mostly academic scientists and the old guard who had served with Kennedy-Johnson—were outraged by the presidential slight. In direct repudiation of the tradition of several decades, he had vetoed a good scientist to what they claimed was a totally apolitical position because of his political leanings. To make the case even clearer, several science societies investigated the possibility of transferring the concept of academic freedom to the field of science advising. Put simply, they wished that federal advisors retained their governmental advising positions even if they led public protests against presidential decisions.

They were being disingenuous. It was wrong, of course, to designate the NSF directorship a nonpolitical post. The job entailed going to Congress to plead for appropriations, and under Congress's new edict the director yearly had to justify what NSF money went to fund. Many in Congress saw the NSF director as Congress's representative in any science policy process. He was to be part of congressional planning. Many congressional Republicans recognized this and also opposed Long's candidacy.[38]

Those already opposed to Nixon took Long's rejection as symbolic, as "the precipitating cause for the sharp drop of scientific confidence in government," and they predicted that few qualified scientists would work in government under those political strictures. Under Johnson, scientists had agitated against Vietnam, classified research on university campuses, and the ABMs. These threads continued and, in addition to the NSF debacle, were joined by objections that HEW investigated the backgrounds of scientists it sought to appoint to advisory panels and likely eliminated those with politically liberal credentials, and that the new deputy OST director had labeled the peer-review system as having "great potential for misuse."[39]

Nixon ultimately appointed an NSF director without any discernable views on the ABMs and politics generally. During this period of turmoil, scientists criticized DuBridge as inadequately representing their interests. Yet he had actively backed increased funding for basic research—the White House supported far larger NSF budgets than the Democratic Congress was willing to authorize—worked to gain Nixon's support of a treaty banning chemical and biological warfare, convinced the president to adopt a greater presence for NASA, and argued strenuously that the Mansfield Amendment would disrupt American science. He counseled his detractors to remember "science and technology are no longer separable from political and social problems." Scientists "frequently find themselves engaged in political discussion and activities," he noted, and concluded that when we enter the political arena we ought not be "surprised if [politicians] judge us on the basis of our political opinions rather than the basis of our scientific competence." Today "science is in politics and politics is in science."[40]

SCIENCE AND SOCIETY

DuBridge tired of juggling the competing interests and resigned a bit more than a year after accepting the post. His tenure in office suggests that the dispute over science was less about money than respect; scientists felt challenged and underappreciated within the new administration. Nixon's nominee, Edward E. David Jr., did nothing to calm their fears. In his mid-forties, David came from industry—Bell Labs—rather than academia, was more of an engineer than a scientist, had no experience advising presidents, and had strong connections to defense industries. A deeply religious man, David dismissed the idea that science could solve all of America's problems. Playing off the popular refrain that if the nation could land a man on the moon, it surely could clean up the environment, provide mass transportation, and cheap healthcare, he argued, "We can't do everything. No master plan will carry us to Utopia." Governance in the modern world was a "series of approximations," trade-offs between options and negotiations among constituencies.[41]

David's appointment produced the usual outpouring of unsettling statements among scientists. Yet the White House was the least of science's problems. Under the guise of balancing the budget, Congress became the primary challenger to federally supported science. Among its biggest complaints was the disruption of college campuses. Protests, riots, and a general lack of order, ostensibly about Vietnam but often encompassing student control of higher education, led congressmen to debate whether the extensive funding sent to colleges and universities served the public well. Proposals generally called for the withdrawal of funding until campus administrators could guarantee proper use of moneys, which in effect meant when student demonstrations ceased. Since much of academic science depended on governmental support, the measure would have stalled ongoing programs and prevented the establishment of new ones.

Of more sustained concern was the contention offered in Congress and elsewhere that science was the cause of many contemporary social problems. Here government support of science seemed particularly culpable. Was government part of the problem? Did government money indirectly go to fund environmental pollution, tainted beef, cancer-producing food additives, and dangerous drugs? Did governmental science lack a social conscience? Did government-financed scientists pursue their own agendas at public expense? What of government scientists per se? Did regulatory scientists regulate the pharmaceutical, food processing, food producing, and agrichemical industries to the betterment of consumers, or did they simply protect the industries that they were charged with regulating? Several different subcommittees investigated these and other questions. Muskie's Senate Subcommittee on Intergovernmental Relations and L. H. Fountain's House Governmental Operations Subcommittee were particularly active. Investigating such diverse topics as regulatory policies of the FDA, chemicals, and the future of man, they called witness after witness outside of government as well as government officials. As in the case of most committees, the congressional efforts became forums for their sponsors' contentions and positions. The Muskie committee, for instance, called witnesses that complained that the nation had long been overconfident about its technological prowess, that this conceit had caused considerable ill-health or other damage, and they demanded government establish a two-pronged attack. It needed to redress the contemporary problems caused by application of technologies in the past and devise for the future means to measure effectiveness of technologies and their social costs before they were introduced. This nascent form of risk-benefit analysis contended that consequences could be anticipated and mitigated. Sometimes working with physical and biological scientists and engineers, research social scientists could draw up normative laws to govern the introduction and utilization of technologies and thus protect future generations. Quite often these calls included a desire to establish the federal government as the arbiter of this assessment, a new mechanism for assessing technologies before they were introduced.

Two major pieces of legislation stand as exemplars of this thrust. Passed in 1969, the National Environmental Policy Act created a new bureaucracy, the Council on Environmental Quality, to represent the environment in governmental decisions. The council was an executive department, its three members selected by the president. Congress demanded that council members be "exceptionally well-qualified to analyze and interpret environmental trends and information" by "training, experience, and attainments." It envisioned the council as an oversight agency, detailing and detecting possible environmental problems before they occurred and pointing out already extant difficulties. An annual environmental quality report submitted to Congress was its mechanism. But the real teeth of the law rested in its requirements for federal departments. It mandated environmental concerns be factored into every governmental action. And it did so in accordance with the Nixon "realpolitik" governance philosophy, which insisted that everything had its costs and benefits, and that costs and benefits needed to be weighed against each other to determine in that instance proper federal action. In that framework, then, the environment was not necessarily more important than job creation or other economic situations. It simply needed to be measured on a case-by-case, instance-by-instance basis.[42]

Social sciences figured prominently in these specifications. Under the act federal agencies must "utilize a systematic interdisciplinary approach which will insure [sic] the integrated use of the natural and social sciences and the environmental design arts in planning and in decision-making, which may have an impact on man's environment." It also stipulated that agencies develop methodologies to ensure that "presently unqualified environmental amenities and values may be given appropriate consideration in decision-making along with economic and technical considerations." These methodologies had to be used in each instance involving the environment in preparing a publicly available "detailed statement setting forth such considerations as the environmental effects expected and the available alternatives to the proposed course of action."[43]

Six months later Nixon used an executive order to create the Environmental Protection Agency. This order pulled "together into one agency a variety of research, monitoring, standard-setting, and enforcement activities now scattered through" government. The transfer permitted "a coordinated attack on the pollutants which debase the air we breathe, the water we drink, and the land that grows our food." As important as the perception that environmental issues cut across departmental lines was the understanding that environment protection might well be at odds with an agency's primary mission. This conflict of interest would hamper activity and give the impression of biased treatment. Left to its own devices, the EPA would establish baselines and methods of measurement as well as gather information. It would arrest and repair environmental damage and abate pollution.[44]

Sometimes material had to be developed. Again, the agency relied on social scientists. For example, it let out a contract to devise a manual that would assist environmental epidemiologists entering any new area. The contract went to an historian, who sought to develop a normative plan based upon the types of information collected in states and cities as part of the public record over the course of the past century. In that fashion, EPA scientists would have a protocol of how to evaluate present circumstance in light of past history. (See fig. 6.)

The second exemplar translated into legislation in 1972. Since the later 1960s, Congress had considered a means to gain control over American science at work. Social scientists and others offered hundreds of science policy plans to gain control of the future. Failure—the inability to predict and prevent—had become unacceptable. It left Americans at the mercy of those introducing new scientific applications and potentially converted citizens into victims; they were enslaved by technocrats, corporations, or other villains. The application of social science—science—would be used to resolve the problems caused by the applications of other sciences. Popular currents reflected the sense of limits, a situation where anything received by one contingent automatically resulted in less to divide up among others, and reminded people that "small was beautiful." "Appropriate technology"—rightsizing technology—rested control closer to the individual user and matched the use for which it had been created. Inflation—as prices rose, the value of money declined—led to wage and price controls, a huge new experiment in

http://www.epa.gov/history/photos/p12.htm

Fɪɢ. 6

A 1972 photograph shows pulp-mill effluent polluting the Columbia River in Washington State, illustrating just one of the multiple environmental problems that drew increasing public, media, and government attention during this period. The Environmental Protection Agency began operating in 1970, the same year that activists organized the first Earth Day celebration.

the federal government's ability to manage American society. The nation's energy needs escalated dramatically just as access to supply seemed to dwindle.

What Congress wanted was an "early warning system" so that it could constrain technology before problems emerged. Using a cold war metaphor was no accident. Congress supposed that one could assess technology before it was introduced and therefore have "first strike capability" in the war in which science was both social problem and social solution. In that sense, technology assessment mimicked the environmental impact statement established somewhat earlier. Congress felt so strongly about the need for technological assessment that it took the unusual step of holding hearings on the matter around the country. Los Angeles and San Francisco were chosen because California seemed to experience first those technological consequences that would soon envelop the nation. Congress also held hearings in the Ozarks to demonstrate to skeptical countrymen the pervasiveness of the technological dilemma. Business interests, squeezed by wage and price controls and under attack as operating outside the public weal, complained that rather than assess technology, any new governmental agency would simply arrest technology. Some thought it would hamper innovation, while yet others labeled the entire venture "absurd."[45]

Congress would not be deterred. Its ability to manage America's present and technological future, uncertain at best, was compromised further by its members' lack of knowledge about science and technology. It needed a device to provide expert, unbiased information so that Congress could plan rationally for a no-fault future. To be sure, the National Research Council provided Congress with access to technical competence, but the legislature felt the group lacked impartiality because it also worked for the White House. Congress demanded a mechanism it would control.

The Office of Technology Assessment operated under a board comprised of equal numbers of Republican and Democratic congressmen, who hired a director and appointed a scientist- and engineer-dominated technical advisory committee. In spirit, the OTA merged the means of the NSF with the agenda of the EPA. It did virtually no research but contracted with a variety of sources—many of who had ties to academia and to the social sciences. Its mandate required the agency to identify "existing or probable impacts of technology," including "cause-and-effect relationships." It also stipulated that the OTA identify alternative technologies to achieve a similar goal, produce a statement making "estimates and comparisons" between the impacts of the various proposals, and identify new areas of inquiry as well as new methods, and report completed analyses to Congress.[46]

The OTA, like the EPA, sought to generate information—knowledge—that policy makers could use to make decisions. At first glance, creation of these two bodies suggested that the problem at hand was a lack of information and that the goal of the two new bodies was to provide that information. That was not the case. It was not that Congress, the president, or the public lacked information per se. They often were swimming in data. What had been the purpose of the EPA and the OTA was to create knowledge its proponents thought they could trust. In the later 1960s and the early 1970s, knowledge was viewed with suspicion; it was merely the expression of the biases of those who created it. The only way to ensure unbiased knowledge was to create it under one's auspices. But that begged the question. It meant that knowledge you produced and relied upon would be viewed by others as unacceptable and biased. Knowledge, especially scientific or technical knowledge, then became an inadequate means to adjudicate disputes. Since facts were not absolute but manifestations of bias, they could not suffice to decide or even help to decide most issues. Without agreed upon facts, the nonquantifiable emotions, feelings, or comforts carried the day.

Governmental regulation was thankless within that milieu. Statements, determinations, and pronouncements all were subject to dispute. Ironically, those regulatory scientists with the most expertise—trained in the field, experienced by years of service in the field—were most suspect since they gained their experience with the interests that they were called upon to regulate. Scientists claiming to serve in the public interest outside of government tended to be young, inexperienced, and, according to many, willing to find fault everywhere to justify their status as public agents. These young men and women quickly erected institutions, such as the Natural Resources Defense Council, Scientists' Institute for Public Information, and the Center for Science in the Public Interest—one group thought about calling itself the Academy of Unrepresented Interests —through which they could issue their pronouncements, lobby legislatures, and work in tandem with lawyers through courts. They also strove to demonstrate to potential donors—these groups relied primarily on donations from persons sympathetic to their message—their centrality to protecting America's consumers. For example, Congress and segments of these groups chided the FDA for not testing drugs to guarantee efficacy and at the same time slammed the agency for the long period of preapproval testing. Despite the passion aroused by pharmaceutical companies, regulation, especially with regard to cancer, proved the most animating single issue. Cancer stood as the nexus around which water, air, and food pollution revolved. The purpose of the EPA was more than simply to keep birds singing à la Carson. One of the most dramatic effects of environmental pollution was increased cancer rates; it was estimated in 1971 that nearly one in three Americans would develop cancer sometime during their lifetime. Cancer so dominated American rhetoric and consciousness that Nixon in 1971 proposed a war to cure cancer as the next great quest for the country that had "split the atom and landed a man on the moon." To spur American efforts, he even tried to get the Soviet Union to wage a friendly competition to see which nation would first conquer the disease.[47]

In FDA regulatory matters, the drug diethylstilbestrol (DES), used in cattle to promote rapid growth, drew the most controversy. Preliminary tests indicated that some DES remained in a few slaughtered animals and therefore ran afoul of the prohibition against adding any substance to food that caused cancer in any dosage in animals or humans. As a prelude to ending DES use in cattle, the FDA by law was required to grant a thirty-day period for comment. Since there was no evidence that DES caused an imminent health hazard, it could not ban the substance immediately. In fact, no evidence existed at all that DES even caused cancer, outside of the fact that it was an estrogenic substance and all estrogens in high doses produced cancer.

Estrogens are female hormones, of course, and therefore found naturally in humans. They also naturally occur in many foods—peas, breads, and salad greens. Dosages of naturally occurring estrogens in peas and bread, for example, were significantly higher than dosages of DES found in those few cattle—parts per billion in the livers and kidneys of less than half of 1 percent of animals examined. More significant, perhaps, was lack of understanding about the relationship between a substance and cancer. Did dosage matter? What would one atom of a substance do? Was there a threshold, a place in which dosages smaller did not matter? Was dosage cumulative over a lifetime? Did various carcinogens have synergistic relationships with other substances? What exactly was a carcinogen? If a substance only produced tumors in the presence of viruses or other biological or mechanical conditions, was the substance truly a carcinogen? After all, it did not cause cancer; it merely permitted cancer to develop in a tainted or damaged environment. How did the unique biology of every individual matter?

Each of the preceding issues was unknown, the matter of conjecture. That hardly stopped critics from demanding that the FDA take drastic action immediately. The Natural Resources Defense Council complained vehemently. Congressman Fountain complained that the FDA's failure to ban the drug led to "wide distrust of government." This was especially so since "the American people, acting through Congress, have declared total war on that most dreaded enemy—cancer." Another congressman warned the FDA to move immediately or the legislature would replace it with "an agency that will." The *Washington Post* reported that the FDA's action showed that the administration's talk about "the conquest of cancer" lacked seriousness. "Who regulates the regulators?" it wanted to know.

Individual citizens wrote the FDA to complain. They disliked their "guinea pig status" and contended that unless DES was banned, "vegetarians will inherit the earth." One woman maintained that "DES may be the cause of feminizing" human males and reminded the male-dominated FDA that "men even old ones are dead frightened of IMPOTENCE." Another woman objected to "being fed poisons . . . just to line some male chauvinist pigs' pockets." Yet others saw the situation as having "reached the national disaster stage" and that "deliberate pollution" of food substances has become "our no. 1 problem." One put the matter succinctly. DES has "done more damage to America than LSD, etc. can ever do."

Animal scientists disagreed. Banning DES was "scientifically unsound in light of current knowledge." But *Science* claimed just the opposite. Arguing that the drug "was a spectacularly dangerous carcinogen," it contended that politics caused the agency not to prohibit use; it would cause the price of beef to rise and stoke inflation. *Science* also challenged the contention of the National Academy of Sciences, the nation's premier science body, which concurred with the animal scientists, as biased in favor of food and chemical companies. Samuel Epstein, a professor of environmental medicine at Case-Western Reserve, put the matter succinctly. "In this country," noted Epstein, "you can buy the data you want to support your case."

Livestock producers joined the fray. They railed against a "very vocal minority," "environmentalists and extreme consumer advocates," "Nader-type groups," "consumer agitators." Some Middle Americans backed them up. They decried the DES matter as "another of many spawned by the disaster lobbies, and hysterics mongers," who were "systematically bankrupting our country and our citizenry of the standards of living which our country is capable of providing and has provided."[48]

The FDA banned the substance after thirty days and the prohibition did not hold up in court. Nearly seven more years remained before the FDA could comply with the various criteria to make its ban stick. For that entire time (and today), complaints raged but little scientific knowledge emerged to help decide the question. But the ban was not about science. Then as now, there exists no scientific evidence of whether DES as a cattle supplement was harmful. What there was and remained was relative certainty—the certainty of assumption and bias—the means by which many regulatory matters were negotiated in the 1970s and after. Newspapers, magazines, and other media were full of DES stories. Those advocating a particular position were understandable. They attempted to manipulate public opinion. But it was those media that tried to be fair, to represent the entire spectrum so that reasonable persons could make informed decisions that were the most meaningful and not for the reasons perhaps that they supposed. To dramatize the distinctions between the various positions, journalists established the boundaries of disagreement; they contacted and featured those individuals asserting the most extreme views.

That tactic galvanized the situation. It made it a matter of catastrophe or harmlessness, perhaps even benefit. Persons were left with only the choice between two extremes. In this fight between disaster and happiness, between heaven and hell, there existed no compromise, only victors and victims. With science unable to contribute decisively, perhaps even meaningfully, the making of policy was nothing more than a glorified, emotion-filled cost-benefit analysis, perhaps also a risk-benefit analysis in which the future of humankind was often said to be at stake. And that emotional analysis occurred without any sort of agreement about what the costs, benefits, or risks were. Who was the victim and who was the victor depended on who was a more draconian political marketer.

Many regulatory agency critics wanted a science in the public interest. When it authorized the NSF to explore and fund applied research, Congress thought it had moved in that direction. The NSF quickly established two projects: Interdisciplinary Research Relevant to the Problems of Our Society (IRRPOS) and Research Applied to National Needs (RANN). IRRPOS gave awards by evaluating "the potential societal impact of the anticipated research and its dependence on an interdisciplinary approach." Environmental matters—pollution and urban blight—captured early awards as IRRPOS worked to add industrial scientists to its usual university clients. One of the earliest recipients was a group of investigators at the Oak Ridge National Laboratory. Establishing an NSF venture within an Atomic Energy Commission entity was unprecedented. RANN's focus was entirely problem oriented, yet it targeted roughly 40 percent of each grant for basic research; the basic research must provide fundamental knowledge useful to understanding or resolving the problem. Here the goal, not scientific curiosity, set the agenda.[49]

Both these new programs added new elements and techniques to the NSF mix. Nixon's NSF budget requests recognized that they took money away from traditional concerns and he asked for about an additional 8 percent yearly to more than cover the costs. At the same time, he asked for less money for military research and drastically sliced NASA's budget. Priorities in Washington had changed. The era driven by nuclear weaponry and lunar landing had passed. Those who expected the status quo now saw "their" money going to environmental, health, and energy questions.

Both because of this shift in administration priorities and because Nixon thought science and technology vital to solving America's social questions, the president took the unprecedented step of delivering a special message to Congress on science and technology. As a good Republican, Nixon saw science and technology as a means to strengthen the economy. "Innovation is essential to improving our economic productivity," he argued, and this was critical because "other countries are rapidly moving upward on the scientific and technological ladder." It was time to capitalize on "the enormous investments which both the federal government and private enterprise made in research and development in recent years." He asserted that "all departments and agencies of the federal government will continue to support basic research" simply for the reason that this research "can help provide a broader range of future development options." To that end, he reoriented NASA to focus on domestic needs—weather forecasting, natural resource exploration, and communications. He demanded that federal standards and regulations reflect the newest, best scientific and technological knowledge and techniques. He authorized the NSF to award applied research grants directly to industries to further the work. He announced new measures to make government patents available for licensing to private enterprise to further job creation and to state and local governments to assist them in resolving environmental and other social issues. He required the NSF to establish a funding program to "support assessments and studies focused specifically on barriers to technological innovation and on the consequences of adopting alternative Federal policies which would reduce or eliminate these barriers." Through these and other similar programs, Nixon wished to create a "new partnership in science and technology—one which brings together the federal government, private enterprise, State and local governments, and our universities and research centers in a coordinated, cooperative effort to serve the national interest."[50]

Managing the Present

Nixon's program neatly summarized his notion of how to achieve stability, prosperity, and order. It depended on management. But Nixon did not want to manage people; he wanted to manage organizations. His entire administration was nothing so much as an organizational strategy. The vast majority of persons he appointed to high-level posts or sought out for advice were trained or schooled in organizational management or had managed an entity—corporation, university, or department—themselves. This was even true of his national security advisor and later secretary of state Henry Kissinger. Kissinger's scholarship had focused on Prince Metternich, whose career established social change as a consequence of order and maintained order by managing relationships between competing foreign powers. To men and women in the Nixon administration, organizational management fostered the order essential for social change. Everything was channeled to the top. Departments and agencies only enacted policy and enforced law. They had no role in policy formation, but when asked for from above they did provide their managers with facts and options. This constituted the planning function within the administration. But planning was propositions, proposals, not maps for the future. What truly mattered was policy and that came from Nixon and his immediate aides. A domestic council—successively Daniel Patrick Moynihan, John Ehrlichmann, and George Schultz—national security advisor, chief of staff, the extremely powerful (because it served as a reality check) Office of Management and Budget—Roy Ash—and perhaps a few others set policy (often argued in cost-benefit terms with political calculations expressly involved) with Nixon. Policy depended on facts and planned-out options, which is what they received from the groups beneath them in the hierarchy. To the policy contingent, there was no such thing as science policy. There was domestic policy and foreign policy and science was absolutely central to both. In foreign policy, for example, science and technology resulted in military wares. But scientists and engineers were also a critical part of détente, a managerial strategy by which Nixon hoped to produce stability by opening venues with China and the Soviet Union. Cultural and scientific and technological exchanges were repeatedly hailed as investments in better relations and a calming of tensions. Science and technology were only one aspect of détente, not necessarily any more influential than ballet, for example, and that sense of multi-dimensionality carried over to the social sphere. Science and technology and scientists and engineers could offer useful material for domestic issues but they could resolve nothing themselves. Domestic issues were an amalgam of economic, social, environmental, and political concerns as well as scientific and technological. All were essential in this schema to the public weal and all needed to be addressed and implemented in a systematic consistent manner to effect social change.

This was the guts of the Nixon White House. He did not bother himself with the affairs of departments and agencies. That was what their managers were for. He restricted himself to the big picture, to establishing policy, and to identifying those units and techniques useful or necessary to carry it out.

Nixon's dramatic reconceptualization of science in government led to some discussion about whether scientists and engineers could be convinced to move their research in these more structured directions. That concern proved baseless. Scientists had grown accustomed to and dependent on federal funding. Appointment of a new NSF director in March 1972 did raise hopes a bit. Stever left Carnegie Mellon to head the agency. His scientific reputation buoyed scientists, and his political affiliation, managerial career, and history on the Nixon transition team impressed the president. His position became even more critical when in late 1972 David resigned. Nixon took the opportunity to refashion the Executive Office to reflect his managerial assumptions. He created new posts of assistants to the president for domestic affairs, economic affairs, foreign affairs, executive management, and White House operations, but his most drastic maneuver was to abolish the Office of Science and Technology within the Executive Office. As a principle, he did not want the Executive Office to "be encumbered with the task of managing or administrating programs which can be run move effectively by the departments and agencies" and transferred the job of science advisor to NSF chief, Stever. Nixon also designated him presidential representative in international scientific affairs. But NSF did not receive the job of advising the Defense Department on weaponry, which now was handled entirely intradepartmentally.[51]

Science furiously claimed that science and technology had been "degraded" by the White House. Abelson put it baldly. "The dismantling of the academic scientific research establishment continues," he complained. "Once scientists were regarded as supermen," but today we "are regarded as mortals." Now, our "views are discounted just as those of any other group." Congress promised to hold hearings on the Nixon reorganization and what that meant for American science and policy making. Presidents of the leading scientific societies, which represented more than 300,000 members, gathered to oppose the reorganization, promising to take a more active role in federal science policy. The situation seemed so dispiriting that the National Academy of Sciences decided a few months later to sponsor a panel of former science advisors to investigate the new White House organization. The panel claimed that the new system was "inherently unworkable" and called for Congress to create a new Council on Science and Technology (CST) to be situated in the Executive Office. This council would be patterned after the Council on Economic Advisers and presumably would assist the president in setting science policy. By law, a CST member would sit as a member of the domestic council and another would sit on the National Security Council. Such an arrangement was necessary because "the presidency requires accessibility of scientific, technological, and engineering counsel" to "decide policy in a fully informed manner." The panel's objections to the Nixon reorganization were telling. Locating the science advisor outside the Executive Office—in this case, a few blocks away—meant that science could not be a full partner in policy making. It also contended that the NSF lacked authority to "maintain discipline among the federal agencies."[52]

Whether both of these issues were accurate or not, they were beside the point. In the Nixon administration, science and technology figured in virtually everything but not in the way with which the science establishment was familiar. Policy making, moreover, was not a single rationalized set of premises, but rather a relativistic competition among options; there was not a search for the best way but rather within the given parameters of economics and politics a cost-benefit negotiated competition for the best possible way to accomplish a goal.

Stever summed up the new science advisory situation when he responded to the NAS panel's report and Congress's concern. Although he had only had the two jobs for about eighteen months, he found the situation eminently satisfactory. That he only saw Nixon at state occasions did not concern him. He had "strong and smooth relationships" with all the White House policy makers, the persons with whom Nixon made the final policy determinations. Within the NSF, Stever had established two new policy units: the Office of Energy Policy and the Science and Technology Policy Office. Both were supplying White House policy makers' analytical reports and advice. In fact, Stever claimed that the two new NSF policy offices together had a staff the size of the old OST and were spending four times as much on policy studies. Both tapped the whole scientific community and therefore were unlike the OST, which, as an outgrowth of the cold war, relied extensively on a small cadre of physical scientists adept at nuclear weaponry. Now the units contacted scientific societies and industrial research groups representing nearly a million scientists and engineers on an ad hoc basis, held meetings, and garnered a wide range of opinion and expertise. In this way, scientists and engineers could be accessed as needed to fit the particular problem at hand.

To Stever, the new arrangement reflected the new vision for and the responsibilities of government. "The issues we're involved in now—energy, environment, food supply—are entirely different" than atomic weapons and lunar landings. As a result, science and technology have a new relationship to policy. They "are only one component along with economics, politics, and social factors." When the president looks at energy, for instance, "he wants a series of options that include every input." Then he and his closest advisors put "together the whole story." Stever concluded by arguing that all this concern about the structure of federal science advising obscured the substance of the matter. A governmental science advisor would be useful for the quality of the advice he provided, his "first and most important role."[53]

Stever's mention of energy was no accident. Energy policy was an area that had gained Nixon's attention. Both Congress and the White House worried around 1970 about a forthcoming fuel shortage. While some proclaimed that "small is beautiful," few in government proposed a less dramatic increase in energy use as a means to lessen the problem. Cutting back on energy might hamper an already fragile economy or lower the quality of American life. American coal proved plentiful but contained high levels of sulfur, which produced acid rain and other environmental pollutants. The lower sulfur coal of the West offered a different dilemma. It rested near the surface and the most effective means to access it was through strip mining. This open-mining technique resulted in a visible chasm of debris and coal-contaminated waste, an environmental eyesore. Efforts to repair the land made coal acquisition costs prohibitive. Natural gas was cleaner and in 1971 Nixon opted to support coal gasification projects as a means to resolve the energy morass. But others contended that reliance on fossil fuels promised an environmental scenario too frightening to ignore. Massive amounts of carbon dioxide unleashed into the atmosphere would precipitate dramatic global cooling and a new terrible ice age. To many politicos and scientists, salvation rested in nuclear power. But there was a problem: conventional light water uranium nuclear reactors were expending uranium at such a rapid rate that the world's supply would last but a few decades. Liquid metal fast breeder reactors would both provide energy and enough fuel for centuries. Nixon also favored this solution, arguing that using fast reactors would be clean energy, lessening pollution of water and air. Noting that development costs might approach $8 billion, the president proposed a government-industry partnership where federal money would pay $2 billion.

Critics dismissed the fast breeder as dangerous. Producing plutonium was itself an environmental hazard. The radioactive material needed to be stored but could leak into water supplies. Plutonium from a breeder could be stolen and used to fashion nuclear weapons. The systems needed a better backup cooling mechanism; radioactive steam could leak or explode from its containment vessel.

America's energy needs seemed so critical that Nixon addressed Congress or the public at least five times on energy matters. But the situation changed markedly in early October 1973. An oil embargo by the countries of the Persian Gulf threatened Americans immediately and in the long run. Quickly, automobiles lined up at pumps as gas stations ran out of gas. Rationing occurred. Doubts existed about the ability of American producers to generate enough heating oil to get through a difficult winter. To counteract this menace, Nixon announced Project Independence, a drive to create American energy self-sufficiency by 1980. Budgets for energy experimentation skyrocketed.

At a gathering of scientists shortly after the embargo, Nixon outlined America's choices. He argued that he favored a clean environment but also that energy was necessary for America's economic future. As a consequence, he challenged scientists to concentrate on the issue, maintaining that in the meantime if he had to decide between "clean air, a better environment" and "freez[ing] to death," he would opt for dirty air because if you are dead "it doesn't make any difference whether the air is clean or dirty." To alleviate what was an estimated 17 percent energy shortage, Nixon asked Congress to suspend clean air standards and waive rules designed to ensure nuclear power plant safety. Before Congress took action, the boycott abated. Although it lasted only two months, energy costs quadrupled. The Atomic Energy Commission continued with its breeder reactor plans and produced a study dismissing the environmental effects of these machines. The EPA strongly disagreed and cited the report as "inadequate" as an environmental impact statement. Matters headed to court.[54]

Before the matter was adjudicated, the Watergate scandal forced Nixon to resign. Gerald Ford, appointed vice president when Spiro Agnew was convicted of accepting money improperly, assumed the presidency. His view of science and technology would differ considerably from his predecessor's.

MARKING TIME: GERALD FORD

Nixon and Johnson had transformed the American scientific establishment and its federal interface. By redefining the federal agenda, Johnson broadened the pool of public claimants for federal authority, extending to health, engineering, and the social sciences the same kind of options and opportunities long held by physicists. Nixon took redefinition to the next level. By conceiving of science and technology as integral elements in virtually every major problem, he fully diffused science and technology within almost every federal endeavor. Through creation of a series of new regulatory agencies, for instance, Nixon placed scientific judgment and technique as a prime adjudicator of the public weal. His appointment of Republican scientists and later his dismantling of the special White House science operation reflected a reality obvious since the 1964 election. The vaunted scientific community was not a community at all but rather a series of diverse stakeholders, united only in their demand for federal largesse and entitlement. Courting primarily Republican scientists, Nixon also increased the number of universities and organizations participating in federal science and technology leadership, thereby undercutting vested interests. Institutional diversity infused new perspectives into the federal mix.

In this context, the numerous and varied calls for a national science policy often sought the status quo ante; in the guise of providing stability, planning advocates often aimed to reestablish models that the mid- and later 1960s and early 1970s had put asunder. Nostalgia for the apparent certainty of the cold war era fueled these programs. But then again what Johnson and Nixon did showed why a national policy as a series of principles to guide future action was fruitless, wasteful, or nonsensical, at best a pipe dream. Both presidents responded to perceived constituencies or problems and each effort formulated relied on science for salvation or amelioration. To attempt to constrain possibilities before issues were even articulated guaranteed failure. It risked erecting edifices irrelevant, pernicious, or wasteful. Each president also stamped his own personality on his administration; he determined not simply what to do—what the "problems" were—but how to do it. This was an issue far beyond style. It revolved around theories of governance. Policy as conceived by its proponents simply sought to constrain presidents from instituting their agenda and substituting with the force of law the agenda of the group or interested entity fobbing off the idea of policy. Policy was certainty, planning, in a world chaotic and not predictable.

Congress had relatively little to do with science policy. This was true even after Watergate with its corresponding congressional awareness that the legislature needed to take back powers usurped by the White House. To be sure, it held the titular power of the purse and therefore should have emerged as a great force. And it needed to pass legislation to enact some presidential plans. Yet it remained too fractured and factionalized to be an effective legislative agency. It was the repository for the criticisms of various stakeholders— "public interest" groups, scientists, economically interested entities, and others, none of whom in their agendas remotely approached any other partisan. As important, Congress rarely moved expeditiously. It championed free and full disclosure, a technique that was particularly ineffective in dealing with new, complicated initiatives undertaken by dynamic presidents to counter ad hoc situations. Rather than a body to shape policies or even vet them, Congress beginning in the 1960s marginalized itself; it only worked at the boundaries, to smooth off the raw edges of a president's agenda.

In truth, science policy in the 1970s did not exist outside of each president's agenda. The long-standing consensus that science

deserved considerable federal support remained, but what that support was and how it was implemented and for what programs
clearly became a presidential prerogative even if law was not drafted
that way. Nixon's refusal to spend moneys budgeted by Congress
demonstrated the potent ability of executive leadership. Congress
was left with little recourse. It could sue—if a majority would
concur—but adjudication would be forthcoming only after lengthy
litigation. And what if Congress lost? Would that not expand the
powers of the presidency even further? Elections could settle the
matter, of course, but those events often proved ambiguous. As
important, the veto emerged as a huge executive bludgeon. Getting
two-thirds of both houses to override a presidential veto proved difficult no matter the Congress's composition. Congress found it more
expedient to work with the president to fashion legislation that
would not be vetoed, which usually meant that Congress resigned
itself to receiving the mere crumbs of authority.

The issues of the mid-1970s and after usually worked to that
end. Ford, Carter, and Reagan were less immediately and directly
concerned with solving America's social problems than resurrecting
its economy. A robust economy meant plentiful jobs—good jobs—
and a reduction of economic-based social dislocations. Yet that did
not mean dismantlement of Nixon's and Johnson's efforts. Programs once enacted proved difficult to terminate. Therefore, many—
most—of the programs seeking to solve social problems erected
during the Johnson and Nixon administrations remained to gobble
up federal moneys. They continued to employ persons in the federal
bureaucracy and provided federal largesse to others outside on a
contract basis. In that way, social science once established remained
vital even as presidential attention turned dramatically away from
that sector. While programs might get marginally smaller, they
tended to remain extant.

When Ford assumed the presidency following Nixon's resignation, he became the first president who had not stood for national
election. In an attempt to get beyond Watergate, he pardoned his
predecessor of all wrongdoing while in office before Nixon even was
charged with any crime. That widely unpopular move was overshad-

owed by Vietnam, where the South's collapse would soon expose the charade of Vietnamization. Inheritance of a persistent energy shortage, heightened by the previous year's Arab oil embargo and coupled with dire predictions of dramatically higher energy usage in the coming decade, a rapidly increasing inflationary spiral and a stagnant economy set a rather desultory tenor for his administration. Ford adopted the rhetoric and manner of a president overwhelmed by the economic and emotional morass. In his first State of the Union address, he began with these staggering facts: "the state of the Union is not good: Millions of Americans are out of work. Recession and inflation are eroding the money of millions more. Prices are too high, and sales are too slow. This year's Federal deficit will be about $30 billion; next year's probably $45 billion. The national debt will rise to more than $500 billion. Our plant capacity and productivity are not increasing fast enough. We depend on others for essential energy."

To the staunchly traditional Ford, two issues most troubled him. First, dependence on foreign energy meant that Americans no longer controlled their future. Only by freeing themselves from the yoke of foreign oil could America maintain its integrity and world primacy. Second, he worried about inflation. This "tax" on all Americans seemed to stem from energy costs and government debt. Each time government spent more than it received, it needed to pay interest on that debt, which in turn increased the amount of money expended by the government. To Ford, this was part and parcel of the dangerous inflationary spiral. "To hold down the cost of living, we must hold down the cost of government," he maintained.

With those principles in mind, Ford set about to put the nation's economy in order. To lessen the impact of inflation, which pushed the poorest Americans into higher tax brackets, Ford proposed a modest tax cut. But he also pledged to hold the line on government spending, with the exception of energy and the military. To reduce dependence on foreign oil, he wanted increased drilling in Alaska and on the continental shelf, strip mining in some circumstances, energy conservation measures in buildings, a moratorium on new clean air standards, more fuel efficient vehicles, development of synthetic fuels, and a major expansion of nuclear power. Ford envisioned within the next decade "200 major nuclear power plants; 250 major new coal mines; 150 major coal-fired power plants; 30 major new [oil] refineries; 20 major new synthetic fuel plants; the drilling of many thousands of new oil wells; the insulation of 18 million homes; and the manufacturing and the sale of millions of new automobiles, trucks, and buses that use much less fuel."[55]

Several of these proposals caused consternation, of course. But by and large, scientists and engineers could only be pleased by the new energy initiatives. Ford proposed a budget that raised federal research and development funding by 15 percent, the highest single year increase in a decade. Much of this funding went to the new Energy Research and Development Administration, which attempted to systematize all energy-related research activities. But more important from the point of view of established professional scientific organizations, he reprioritized NSF funding, reducing by 5 percent the money spent on the Research Applied to National Needs (RANN) program, which aimed to devise methods to ameliorate social problems, and expanding the basic research kitty by more than 13 percent. *Science* championed the move, arguing that "the transfer of the White House reins" ended "the antiscience aura of the two previous administrations."[56]

Ford's decision to fortify basic research spending stemmed not from latent fondness for science but rather from a hardheaded practical analysis of what science could do to further his agenda. Indeed, his career as a legislator with respect to science was mixed. In his quarter-century in Congress, Ford voted for measures that would expand science funding as often as not. But never did he waver from a firm commitment that basic research was necessary for the public weal. To Ford, basic research was American jobs once removed. He was sufficiently old-fashioned to believe that studies into the principles of science or of some phenomenon were among the best ways to generate knowledge that would fuel new processes or practices. But this understanding was not simply about new avenues to wealth. It also meant newer, cheaper means to accomplish already standard practices; in addition to opening new economic sectors, science promised to increase American productivity in already-established venues. The latter situation was almost as troublesome as the former. Long-dominant American industries were threatened in world markets by cheaper overseas labor and new production methods. Japanese and German automobiles flooded the American market as did electronics from several Asian countries. Even steel seemed tenuous, on the verge of going the way of the passenger pigeon.

The fact that Japan and Europe had markedly increased basic research funding during the past decade while America lagged behind confirmed to Ford the nexus between basic research and prosperity. He recognized that new, well-paying jobs would reduce inflationary pressures and mitigate if not resolve many of the nation's most persistent social problems. Government would be called on to do less, which left individual citizens more money to invest or use as they might wish. But Ford also understood that investing in basic research was an appropriate governmental responsibility. Governments served citizens and undertook tasks that benefited society but remained outside individual purview. To Ford, for individuals or corporations to invest in basic research was a poor investment. Most basic research ended in failure, dead ends; investors would suffer precipitous losses. Requiring large sums of funds, basic research must not be dependent on the whims and capriciousness of individual decision making. Finally, the knowledge generated from basic research should not be proprietary. Basic research was an investment for society. Its payoffs should extend throughout society.

A similar hardheadedness about government's role drove Ford to increase defense spending. Years of fighting in Vietnam and trying to right America's social wrongs had caused the national defense to deteriorate. Maintaining military strength was to Ford a necessary investment in the future. A strong, technologically savvy military would curtail the menace of foreign adventurism and thus actively decrease war possibilities.

Congress, which represented more diverse constituencies, offered a much less-unified view. Increased military spending remained controversial, even problematic, but nuclear power clearly galvanized interest. Groups in several states, including California, had mobilized against nuclear power. Nader lent his weight to the crusade as well as his organizational skills. A Washington beachhead of lobbyists, supported by acute technical advisors, pledged to stop the nuclear juggernaut and pressed the matter in Congress. Industry representatives challenged these determinations, which highlighted the lack of safety plans for reactors and the worries about storing and protecting plutonium. Arguing instead for clean technologies, antinuclear forces urged Congress to phase out extant plants and phase in clean geothermal and solar technologies and strict conservation measures. The OTA chimed in on the side of the antinuclear forces. Ford's reliance on nuclear power ignored the likely political firestorm, the OTA reported. By focusing only on narrow, high-tech hardware, Ford ignored "incentives for commercial application, environmental constraints, competition for the use of scarce resources, and public resistance," nontechnical issues even more important in the political arena. Practically, this would mean that much, if not most, of Ford's energy funds would be wasted, placed in programs that had no possibility of adoption, and the resulting delay would stymie America's energy prospects for the next quarter century.[57]

The OTA offered a similar critique of coal liquefaction. But Ford's decision to pour money into research and development and into other forms of basic research continued throughout his administration. The budget he proposed for 1976 again held the line on federal spending but provided generous energy, military, and basic research increases. NSF funds rose a whopping 20 percent, much of it outside energy. Ford maintained a consistent policy, but Congress continued beset by conflict. The NSF itself emerged as a cause célèbre. Beginning in March 1975, Senator William Proxmire (D-Wisconsin), powerful head of a subcommittee responsible for NSF activities, began a monthly announcement of the federal grant that most abused public trust. An $84,000 NSF grant to explore why people fell in love received the first Golden Fleece Award and brought attention to the types of research funded by that government agency. But another NSF program the next month set off a bigger maelstrom. The NSF funded an introductory anthropology course for ten-year-olds—Man: A

Course of Study. Purportedly including "adultery, cannibalism, killing female babies and old people, trial marriage and wife-swapping, violent murder, and other abhorrent behavior," NSF approval led some congressmen to question if the NSF did not require more aggressive oversight. To that end, Congressman Robert Bauman (R-Maryland) sponsored a bill compelling the NSF to clear every research grant with Congress before funding could occur. This measure mandating congressional scrutiny passed the House but would fail some months later in the Senate.[58]

The House's action not only placed the agency on a short leash but it also undermined the concept of peer review. Each NSF grant had been reviewed by several scientists, each an expert in that area of study. That the House maintained that its judgment should supercede relevant investigators angered scientists precisely because it further undercut their autonomy. Congressmen sometimes attributed poor decisions made by the NSF not to peer review per se but rather to how the NSF applied peer review. The agency depended on a small number of scientists at a small number of institutions. Instead of representing the profession, congressmen felt that peer review at the NSF was a bastion of undeserved privilege, surely not unbiased and perhaps not honest. Proxmire broadened the critique by claiming that the NSF's "academic oligarchy" favored the universities that they represented when funding was awarded.

The NSF at first declared peer review sacrosanct and refused to provide Congress information about evaluations or reviewers, but a series of summer, 1975 congressional hearings caused the agency to recant. The NSF's oversight board, the National Science Board, revamped NSF procedures. It demanded that the NSF publish a list of reviewers yearly. Publication of the list would guarantee that the agency employed referees broadly representative of American science, a criterion that the NSB noted must be part of the review process. The NSB also mandated that persons who had submitted grants would now receive verbatim copies of all reviews. Finally, the NSF, if asked, would now explain to an unsuccessful applicant on what grounds the grant was refused.

Congressional complaints and others from outside revolved

around issues other than peer review. Critics claimed that the USDA ignored what was becoming a world food crisis and ought to devote more assets to research. Three EPA attorneys protested what they contended was the agency's backsliding on pesticide regulation. An increasingly turbulent weather pattern led some climatologists to complain that federal government was not adequately prepared for what promised to be a weather-related global famine. The *Los Angeles Times* reported that the EPA distorted sulfur emissions reports to prove that they had a harmful health impact. A prominent NIH cancer researcher resigned from a carcinogenesis study because he felt a lack of funding and manpower jeopardized results. A federal nuclear power safety panel member painted present nuclear safety techniques "totally inadequate to the complexity of the problem." An ethics panel decried the NIH's refusal to follow the panel's fetal research guidelines.[59]

Recombinant DNA research both typified the disquietude about the scientific enterprise and broke new ground at the same time. Being able to disengage and recombine genetic material in different ways, including those not naturally occurring, seemed almost like creating new life-forms. Some scientists and others considered biology now at the place of physics with the first atomic bomb. Life scientists lost their innocence; for better or worse, biologists seemed to hold power to affect human destiny.

Unlike World War II physicists, those adept at the new recombinant DNA technology decided in late 1974 to put aside their studies for a bit and consider ways of safeguarding the environment. Under the auspices of the NAS, they met to hammer out a protocol the next year. Setting as their goal to contain these nonnatural organisms, they established perceived categories of risk to restrain the potential biohazards. At the same time, they acknowledged that what they thought to protect against was not nearly as dangerous as those things not foreseen. If such an event occurred, it "will not be like DDT." You cannot "just stop manufacturing" it if it gets into the environment. It will manufacture itself and perhaps become a scourge of humanity, passed from person to person and generation to generation.[60] (See fig. 7.)

http://libraries.mit.edu/archives/exhibits/asilomar/index1.html#2

Fig. 7

Influential biologists Phillip Sharp (left) and David Baltimore (center) engage in discussion during the February 1975 International Conference on Recombinant DNA Molecules. At this meeting, known as the Asilomar Conference, molecular biologists assessed the risks of their work and proposed guidelines for research safety to prevent the accidental release of experimental organisms and any threat to public health or environmental well-being. The self-regulatory rules that the conference formulated proved influential, but did not address the complex and controversial ethical and legal issues surrounding genetic engineering.

The group's general principles were then tapered by each nation's central biological agency to suit research in that place. The NIH held that function in America and announced its preliminary guidelines in late fall 1975. Two distinct criticisms appeared. The vast majority complained that NIH guidelines were too lax and offered insufficient protection against menace. But a very vocal minority argued the other case, that NIH rules were unnecessarily strict and would hamper inquiry. Both sides attributed the brouhaha in part to conflicts of interest within the NIH. New subcommittees to redraft NIH rules were proposed.

At the heart of the matter rested a simple assumption. Regulators did not know what needed regulation and how to ensure that it remained regulated. Two political doomsday scenarios challenged the regulators. First, "mounting impatience" among biologists pressured the committee to end its determination quickly, as did threats of "Saturday-night experiments" where some ruthless scientists took matters into his/her own hands and commenced forbidden work without reasonable safeguards. Second, rules not sufficiently restrictive might convince Congress that scientists were incapable of governing themselves and precipitate draconian legislation severely limiting inquiry.[61]

A few biologists offered a far bolder perspective. They argued that free inquiry was a scientific right and that any attempt to circumscribe the ability to do whatever research they chose was an intellectual abomination. References to the Church's treatment of Galileo often peppered these considerations. More common was the opposition, a pronouncement that society granted scientists the right to research and therefore could determine what was appropriate.

The NIH opened the process to public comment but the harshest reactions remained in communities. New York State debated the wisdom of conducting recombinant DNA research within its borders and Cambridge, Massachusetts, placed a ninety-day moratorium on the work until a select committee of its citizens made recommendations about DNA research within the locality. Both bodies found the NIH an illegitimate regulatory body, arguing that an agency whose primary mission is to sponsor work in a field could not be an effective, impartial regulator of that work. Its interest was vested in the work, not in protecting the public from the work's potential consequences. The public needed to decide the matter. Scientists could provide data and arguments to convince—educate—the public, but the decision rested entirely with the public.

Who represented the public? In the case of Cambridge, Massachusetts, an elected city council assumed that authority. Yet they were only one claimant. "Public interest" organizations also claimed that mantle. On what grounds did they assume that authority? They did so simply by claiming that right, simply by designating themselves representatives of the person or persons upon whom science was or might be plied. By automatically placing themselves as adjudicators of the decisions of others, they in effect erected themselves in opposition, as an antiestablishment establishment. That these public interest organizations might be necessary, or make themselves quite good at what they had claimed for themselves, did not contradict the point of their origin. They simply established themselves in opposition to the status quo and that made them therefore somehow seem both more virtuous and credible. Their status, their position, their acceptance stemmed from their opposition, an assumption rife with a belief that without this opposition all was lost.

"Public interest" groups in effect designated themselves as stakeholders. And they flourished because of their opposition. To agree, to concur, was an invitation to cease to exist, to become irrelevant. They had no independent existence. Their continued vitality depended entirely on their arguments against standard or proposed practice.

Senator Edward M. Kennedy (D-Massachusetts) understood the nature of this adversarial relationship. He had long backed greater funding for scientific efforts in part to overcome omissions or errors noted by public interest organizations. His reputation as a staunch supporter of science received a huge hit, however, when in late 1974 Kennedy began to argue that science was too important to be left to scientists and that a national commission should set the national scientific agenda. Composed of a significant majority of laymen, the commission would have a twofold purpose. As Kennedy gave his proposal a more full articulation during early 1975, the committee would dictate what the public was willing to fund—it would set the federal science program—and it would safeguard the public from the unfortunate consequences of science. In the latter case, he noted the recombinant DNA situation and while he praised scientists for having the good sense to stop and reflect on their work, he decried their decision to consider the issue a scientific matter. To Kennedy, science was too important and dangerous to be left to scientists, and he held a series of Senate health subcommittee hearings to make these points. Witnesses argued that the concept of informed consent should extend to virtually any scientific act, requiring scientists to gain public affirmation before undertaking any experiment, and that simply to use scientific terminology to consider an issue did not make that issue scientific. Overwhelmingly they disagreed with the premise that only scientists had the special knowledge necessary to recognize potential hazards from research and to devise appropriate resolutions.

Kennedy's hearings produced no legislation and he moved in a complementary direction, proposing in late 1975 that a portion of NSF funding go to assist "public interest" organizations. This idea apparently did not come from Kennedy or his staff but rather from Frank von Hippel, a theoretical physicist associated with the Center for Environmental Studies at Princeton and coauthor of *Advice and Dissent*, which claimed to lay out a framework for "bringing technology under democratic control." Relying on "public interest" groups as an effective counterpoise to scientists receiving governmental funds, von Hippel had envisioned a fellowship program for scientists to work with public interest organizations and a journal of public interest in science to address these organizations' concerns and thus to unite them into a more effective opposition force.[62]

Kennedy maintained that his "Science for Citizens" initiative would redress what had become an obvious imbalance. He was joined in this quest by Jacob Javits (R-New York). Worried about this new program's potential ramifications, the NSF attempted to control what might be involved in this new program by holding seven regional forums in December 1975 to gather public sentiment. A scant two hundred persons spoke at these events, each of whom had to submit a written statement outlining its proposal prior to a session.[63]

The NSF's hearings and Kennedy's proposal found considerable

favor in "public interest" circles. One scholar likened it to the Protes-
tant Reformation, contending that the "pope," science, was about to
be replaced by a priesthood of true believers. Others talked of rene-
gotiating a broken contract between the sciences and the public they
were designed to serve. Another complained that the matter was
complicated by the multitude of stakeholders—scientists of many
stripes, "public interest" leaders, philosophers, journalists, and congressional staffers. Finding anything like common ground seemed
extremely difficult. Few offered specific platforms. Labor unions
proved an exception. Their spokesman called for a panel composed
of labor union representatives, welfare recipients, citizens' groups,
and others to identify problems that researchers needed to address
and to evaluate the results of research previously undertaken.

Kennedy's "Science for Citizens" bill passed the Senate but was
rejected by the House. A joint conference committee spent three
months wrangling over that provision before accepting it. Even then,
the chair of the House conferees refused to sign or vote for it. The
law authorized about $1 million in NSF grants for fellowships to
place scientists within "public interest" groups and for forums, conferences, and workshops where scientists could provide technical
advice to those opposing government programs. "Public interest"
organizations saw the measure as a step in the right direction—they
got to educate scientists about the "public interest"—while others
viewed it as creating an entrenched opposition. "Public interest"
groups would have access to technical expertise better to oppose
government pronouncements and programs and to file lawsuits and
protocols on behalf of the "public." But a new situation presented
itself. Now that "public interest" organizations had access to the
same sort of funds that other scientists had—now that they were
supplicants and recipients of federal largesse—could they be
trusted? Did federal funding make a group less reliable, a creature of
a special interest? Was not that the critique of government regulators
and other government-funded scientists? Why would that not be the
case with those that had claimed to be serving the "public interest?"
Was that not the definition and template for government itself?

The Ford administration took no position on the "Science for

Citizens" issue, concentrating instead on efforts to create a new science and technology office within the White House. As soon as Ford had become vice president under Nixon, a few congressmen and representatives of the organization of scientific society presidents called on him to press him to establish a science office if he ever became president. Ford made no promises, but it let it be known that he felt a science office a useful entity.

Ford's support of a science office stemmed from his idea of what science could accomplish within the federal government. Unlike Nixon, Ford refused to recognize science in every federal program. His much more limited federal science vision was reflected in his budget priorities—basic research, defense, and energy. As critical, Ford did not conceive of management and policy formation as had his predecessor. With Ford's considerably more circumscribed federal agenda, there was much less to manage or to balance against another vital program. Ford's desire to reduce markedly the scope and activities of the federal government meant competition among social groups over social issues rarely occurred; much of what government attempted under Nixon was dismissed as outside government's purview under Ford.

Two anecdotes demonstrate the passion of Ford's limited government philosophy. To fight inflation, which Ford recognized as a significant enemy of all Americans, he proposed a voluntary program, marked most incongruently by buttons with the acronym WIN, which stood for Whip Inflation Now. He displayed a similar sense of federal restraint when New York City—in part because of inflationary pressures—teetered toward bankruptcy. Ford refused federal intervention. Next day's *New York Daily News* headline—"Ford to City: Drop Dead"—helped underscore his unshakable commitment to reducing federal intervention in even the most unpopular ways.

Ford saw an executive science office as a place to set the agenda for basic research, energy, and defense, little more. That was not how Congress viewed the matter, however. The House put together a bill to placate scientists, while the Senate attempted to establish a firm, service-oriented science policy for America. Even before Nixon's res-

ignation, Kennedy had demanded a new three-person federal science advisory system. These advisors served both the president and
Congress equally. In this manner, policy advice would be consistent
through the two branches. Part of the task of these advisors was to
make sure that federal research and development spending constituted a mandated percentage of the Gross National Product and that
every dollar spent on military research would be matched by one
spent on nondefense-related science. Much of this nondefense
money would go to a new civilian science systems administration
within the NSF to meet "the human needs of the nation in such priority problem areas as health care, poverty, public safety, pollution,
unemployment, productivity, housing, education, transportation,
nutrition, communications, and energy resources." Each attempt to
resolve these social problems needed to be appraised according to
"technical, environmental, economic, social, and esthetic factors." A
board composed of equal numbers of business leaders, labor representatives, scientists, engineers, social and behavioral scientists,
environmental and community groups, and consumers would dole
out the new agency's funds.[64]

Kennedy's bill articulated in the most intimate details what the
new science office would do and how it must do it. The House
measure was less specific and set a slightly more limited agenda. It
called for thought-out strategies to use science and technology to
reach foreign and domestic policy goals and depended on a five-
member advisory committee to formulate those strategies. This
policy committee would be separate from the various implementation efforts, which would be housed in a new cabinet-level department of research and technology operations. This new department
would oversee the NSF, NASA, the Educational Research and Development Administration, the National Bureau of Standards, and a
new Science and Technology Information Utilization Corporation.
The new secretary would not only set budgets for these agencies but
also other science-based departments, such as the EPA and the FDA.
In that sense, a new science czar would coordinate and rationalize
the application of US science policy, while the advisory committee
would recommend its formulation.

Both plans far exceeded the scope of what Ford wanted and he let Congress know that either would face an immediate veto. That was not to suggest he was without scientific advice, however. Stever, David, William O. Baker, president and chairman of Bell Labs, and Simon Ramo, cofounder of TRW, a defense electronics firm, served as a sort of kitchen cabinet of science advisors. At the House's urging, Ford in May 1975 laid out the type of executive science office he was willing to accept. He envisioned a staff of about ten and a budget of about $1.5 million, which would provide advice about military science as well as domestic policy. Ford's advisor would not be a member of the cabinet, but he had no objection if the Senate wanted the opportunity to confirm his nominee.

Congress had little recourse. It could not force a science office on a president and expect him to seek its wisdom. As important, Ford had the statutory power to fashion an executive science office himself. It was his choice to work through Congress. Almost overnight, then, the House jettisoned its bill and constructed another in line with Ford's desires. Its main difference was that it mandated that the advisor "shall participate through the budget development process" but left it to the president to determine what that phrase meant. The bill also proposed a major study to take stock of the "total context of the federal science and technology effort." Its goal was to stimulate organizational reform, simplify regulations, increase effectiveness of sponsored research, and streamline means of transferring new technologies to the marketplace.

Kennedy and the Senate were less willing to adopt Ford's guidelines per se. But they also recognized that they had little power to force the president to do something against his wishes. As a consequence, the Senate nibbled around the edges. It tried ever so gently to expand the new science office's authority, a fact that led the body to demand that it be called the Office of Science and Technology Policy and that the term "engineering" appear whenever the word "science" did. The Senate bill also pushed for Senate confirmation of deputy advisors and argued that they could be called to testify before Congress but also noted that much of what they did in the Executive Office was privileged communication. The Senate also urged that a second study also be undertaken and include an analysis and forecast of potential scientific and technological problems and to recommend possible legislation. Finally, the Senate wanted the science advisor to coordinate scientific and technological activities among the relevant federal agencies.[65]

None of these modifications proved particularly onerous to the White House and it signaled its assent. In any case, Congress could not reasonably compel the Executive Office to do much of anything. In May 1976 the OSTP was established. Ford picked Stever to be the first director and he was approved in August of that year, just in time for the presidential election.

Chapter 3

FROM MALAISE TO MORNING IN AMERICA

WWJD (What Would Jimmy Do?)

Democratic presidential candidate Jimmy Carter followed in the time-honored tradition of establishing a science advisory team to reflect a candidate's deep commitment to science and therefore to attract support from scientists and others for the campaign. Accepting AAAS and NAS recommendations, Carter appointed university- or business-employed men and women long involved in advising. These people never met as a body but provided the Democratic nominee with policy papers so that he would hit the ground running once elected. To reflect the dynamism implicit in this form, Carter called the assemblage his "science policy task force." But while this group represented mainstream science, it was not Carter's only policy foray. As early as August 1976, several months before the election, he also established a science transition team to identify and select science policy people for the new Carter administration. There Carter ignored the mainstream and went to the "public interest" movement. His Energy and Natural Resources coordinator was director of the Environmental Policy Center, a group committed to eliminating strip mining and outer continental shelf oil exploration, and his Health and Welfare coordinator was legal director and general counsel for the Southern Poverty Law Center. His coordinator for government organization was the first executive director of Nader's Center for the Study of Responsive Law, Harrison Wellford. Wellford cut his teeth as a raider, writing the persuasive *Sowing the Wind: Food Safety and the Chemical Harvest*, which skewered the USDA and the FDA and was later accepted as his Harvard PhD dissertation. Wellford subsequently got a law degree.[1]

Carter's choice of "public interest" lawyers to head his federal science transition team disproved the notion that scientists must frame science policy; it substituted legal maneuvers for scientific determination or judgment. Refusing to abide by established procedure, or shattering the status quo, categorized Carter's entire presidential campaign. He ran as an outsider, parlaying his absence of governmental experience beyond Georgia's borders into a virtue. Dismissing knowledge of and experience in the federal enterprise as inhibiting, constraining, Carter claimed not to be wedded to the methods of the present or beholden to solutions already in place. As critical, he claimed to be free of the interests upon which then-implemented programs were predicated and as if to demonstrate that contention, he erected a governmental system based upon notions not commonly associated with modern government. As a born-again Christian, Carter sought to bring Christ's teaching into everyday life and as a basis for governance. His was a framework in which the principles of Christian fairness and integrity would trump greed and profit. Government would represent the disenfranchised, those discriminated against, the disadvantaged. This sense of a higher calling caused him to apply his moral code as consistently as possible, even to foreign policy. For example, to try to force the Soviet Union to behave as a good world citizen, Carter halted exports of grain to that country and refused to allow American athletes to participate in the 1980 Moscow Olympics. That Carter's decision reduced American farmers' revenue, that America's trade imbalance skyrocketed, or that American athletes lost a once-in-a-lifetime chance to compete in the Olympiad was a small price to pay toward the greater goal of a more responsible Soviet Union.

Carter's Christian morality emphasis attracted him to those persons claiming to apply science in a similar, albeit secular, fashion—those in the "public interest" movement. Morality as posed by Carter was absolute; it brooked no relativistic appraisal. Carter's Christian morality test also negated previous means to set governmental policy and to achieve governmental efficacy. Both the cost-benefit analysis and risk assessment of the 1960s and 1970s had balanced competing factors to achieve what appeared a most salubrious compromise. In that sense, these governmental determinations had reflected contemporary politics in which various stakeholders battled for control and power without an accepted means to adjudicate between the merits of their contentions. In Carter's world of morality, science policy was Christian morality. Since morality guided decision making, the application of science then was merely technical.

Carter's distinction between science policy—Christian morality—and its application accounted for his bifurcated advisory system. His technical advisors had real-world experience; business and university leaders understood how to apply science to achieve precise ends and had the technical savvy to pursue the most efficient, cost-effective means. Carter himself readily identified with this group. His modest business career in Plains, Georgia, stemmed from his scrutinizing every detail of the peanut-growing and distribution enterprise. As the first president since Herbert Hoover to have extensive scientific training, Carter was well versed in science, regarded it as critical, and understood various ways to use it to come to bear on social questions. One scientist quipped that it was about time that America had a president who had taken two years of calculus. But his Christian template restricted Carter to manufacturing those science-based solutions to the issues confronting Americans that were consonant with his morality. Carter's mania for micromanagement guaranteed that systems created to apply these social remedies would both maintain moral orthodoxy and be as lean and focused as possible. To Carter, efficiency per se was empty. Decency, livability, and justice gave it meaning.

Using Christian morality as the standard of governmental action and adjudicator of success meant that Carter applied a measure at odds with most Americans. During a period characterized by the journalist Thomas Wolfe as the "Me Decade," and marked by a Pall Mall quest for individual fulfillment, effectiveness had become personalized. Individuals felt free to evaluate activities within the context of their own personal perspective, asking the critical question: did a particular act further my personal goals? To Carter and other born-agains, this meant applying Christ's lessons to contemporary life. But to the vast majority of Americans, many of whom found themselves rocked by a world of increasingly limited possibilities—the energy crisis, galloping inflation, huge trade debt, a nation seemingly in decline— charity began at home. Individuals expressed outrage over recent events and demanded recompense; increasingly government became the forum for redress. Failure of government to rectify your complaint became synonymous with governmental failure and incompetence.

The vast chasm between the expectations of many Americans and Carter's idea of government's role promised that his presidency would be turbulent. Right after his election, Carter moved to implement his Christian morality plans. He initially appointed twenty-four "public interest" lawyers to high-level posts within his administration, thus providing a clear and consistent moral compass for government action. But Carter's reach extended far beyond lawyers. He selected Carol Tucker Foreman, president of the Consumer Federation of America, as assistant secretary of agriculture for food and consumer services. A move akin to having a fox guard a hen house according to some, Foreman served as the moral conscience to what had traditionally been a producer-oriented agency. His choice of Bob Bergland as USDA secretary demonstrated the same thrust. Bergland had long operated as a spokesperson for rural America, emphasizing collectivism and the role of cooperatives in furthering a vital rural America by fostering small and middle-sized farmers in the wake of a corporate farming onslaught.

Carter also introduced moral test-based policies. Administration members had to agree to scrupulous ethical regulations that forbade them for a period of two years from entering industry or university employment where the activities they had undertaken during government service might be used to create a conflict of interest or an unfair competitive advantage. Scientists especially objected to this ethical construct. A strict construction of this regulation, they complained, "would cut off the government's supply of first-rate scientific talent."[2]

A similar moral sense led Carter to suggest decreasing medical college funding. Struck by an extremely tight federal budget and high inflation, Carter noted that an anticipated glut of doctors virtually mandated that he end the practice of granting medical schools a direct 8 percent premium for every student they enrolled. In effect, Carter reasoned, the federal government was needlessly subsidizing persons for which there was no national need. Besides, graduates would receive impressively high incomes from which they could easily repay any loans taken out to pursue their degrees. Without the 8 percent premium, schools would lack incentive to expand enrollment. The federal money saved could then be used for compelling national necessities.

But perhaps the clearest example of Carter's public interest/ Christian morality/outsider thrust was the selection of Donald Kennedy to head the FDA. A well-regarded neuroscientist, Kennedy had virtually no training in regulation or with food and drugs. He came to the subject out of a deeply felt commitment to public service and almost immediately had parlayed his neurobiological reputation and desire for public service into important positions. Before he had accepted the FDA post, Kennedy had chaired the National Academy of Sciences study on alternatives to pesticide use and joined the World Food and Nutrition Study. Sponsored by the National Research Council, the study tried to determine where scientific research could be best applied to lessen world hunger and enhance world health.

Virtually every constituency that dealt with the FDA praised Kennedy's appointment and for rather similar reasons. They saw him as a man of science, an archchampion of the application of fact-derived knowledge to every problem. Ironically, they held these views even as the FDA operated in a much murkier manner. What was fact and what was a meaningful fact were often up for debate. In previous administrations, cost-benefit and risk-benefit analyses served as relative truths. Under Kennedy, those types of situational ethical compromises were to give way to clear black and white.

Each group favored Kennedy because each thought it had science on its side and Kennedy would in effect rule in their favor. The reality of the situation proved quite different. Consideration of food additives, such as nitrites—used in the preservation of sandwich and other meats—proved problematic but the tumult over saccharin demonstrated the misapprehension of simple adjudication.

Citing Canadian studies on rats that showed a possible correlation between saccharin intake and bladder cancer, Kennedy's FDA proposed a hearing as the first step to banning the artificial sweetener. Almost immediately the agency was attacked on all sides. Some pointed out that the relationship between animal studies and human health were unclear; animals often got different diseases and different rates of similar diseases than humans, so extrapolation could be difficult or even wrong. Others contended that dosage mattered; a human would have to consume at least eight hundred twelve-ounce cans of saccharin soda per day over its lifetime to receive the same relative dosage as given to the Canadian rats. Still others argued that any hint of increased human cancer risk demanded that the FDA immediately ban the substance.

Other claimants, operating outside of Carter's framework, argued as they had in the immediate past, within a cost-benefit framework. For example, the American Cancer Society backed the continued use of saccharin because banning the substance "may cause great harm to many citizens while protecting a theoretical few." Saccharin seemed to offer "potential benefits . . . to diabetics, persons with heart disease, obesity, or other medical problems." The American Diabetes Association demanded "soda without fear" and maintained that diabetics needed saccharin "to enhance their quality of life." A psychiatrist who treated juvenile diabetics contended that snack food was such an essential part of the teen experience that sugar-free varieties were crucial to a young diabetic's "psychosocial development." Increasingly, however, persons questioned if the benefits that saccharin advocates claimed were true. Were there studies that showed that saccharin reduced someone's caloric intake? None seemed readily available. Proponents argued that those studies did not exist because the results were self-evident; every diet drink was one less sugar-loaded drink. Others proposed that sugar lust merely took different forms. Someone might have an extra cookie or two with their calorie-free beverage. And yet another group pointed to one study that suggested that ingestion of saccharin actually lowered blood sugar, which increased appetite and led to obesity "and, among diabetics may actually contribute to the onset of hypoglycemia and insulin shock." Some worried about other groups, especially children and women of childbearing age, who in other instances had shown exquisite sensitivity and toxicity to drugs that barely affected other portions of the American population. Might they be victimized by legal saccharin?

To Kennedy's FDA, the science was clear. Saccharin must be banned. But the Carter administration's framework for action rested outside of most Americans', and through Congress they voiced their opposition. Citing the same arguments as they had to the FDA, which dismissed them because they were generally cost-benefit based, Congress embraced them and prohibited the agency from banning the substance. Saccharin would remain legal. It remains so today. Kennedy put a good face forward when he maintained that even though saccharin continued in use the whole exercise had been "beneficial" as "a good national education on toxicology problems."[3]

Carter himself participated publicly in few regulatory matters, which he took as questions of pure science applied through a moral gauze. He intervened little in the controversy over setting the parameters of recombinant DNA work, which mesmerized many scientists and galvanized Congress, because he was confident that science would prevail and the public interest would be served. Carter focused instead on dramatic questions that science was powerless to resolve. Indeed, he spent much of his presidency attempting to get what he saw as the major issue and threat of the day under control. That issue was the possibility of a nuclear holocaust.

Carter perceived this issue as multifaceted, as inextricably inter-twined with a series of other problems. To be sure, he worried about the potent menace posed by the Soviet Union. But Carter deemed the USSR as not as inherently dangerous as a series of nations with nuclear capabilities. There the danger would be from many places, each with its own formidable agenda and prospects for using the weapons to intimidate foes and to increase its influence. As an established superpower, the Soviet Union controlled much of the world and had nearly as much to lose as the United States in a superpower confrontation.

On the simplest level, Carter worked to reduce the number of nuclear weapons. Patiently and repeatedly, he tried to convince the Soviet Union that their future rested in significant reduction of atomic devices. Carter proposed cuts of the size that would have made Mutually Assured Destruction (MAD) problematic. Even when the SALT II treaty was concluded in 1979, which limited not only the number of nuclear missiles but also the number of warheads, Carter urged Brezhnev to sign a codicil compelling the two superpowers to reduce further their atomic arsenals by 5 percent per year.

Power Politics

The threat of nuclear proliferation fueled Carter's staunch opposi-tion to the breeder reactor. Championed by the nuclear power industry for its ability to produce more fissionable material than it consumed, the breeder promised to solve America's electric power requirements for the foreseeable future. The country would no longer be dependent upon the Soviet Union or the developing world for fissionable fuel but could manufacture as much as it wanted, and even become a fuel-exporting country.

Carter's resistance to breeder reactors stemmed directly from his belief that creation of additional fissionable material was a threat far more potent than energy shortages or price hikes. He warred against the breeder at every turn, refusing to appropriate money Congress authorized and refusing to barter any program for the breeder.

Carter did so at the expense of the American economy. France, England, Japan, and Germany roared ahead in breeder development, even selling reactors to Brazil and other countries. American products might have dominated those markets had Carter given the green light to the breeder.

To fuel American industries and to warm American homes without breeder technology, other, more expensive but less dangerous energy sources became a must. In 1977 Carter proposed massive taxes on carbon dioxide–producing, environmental-polluting oil, gasoline, and natural gas, both to raise revenues and to limit usage. Calling his program "the moral equivalent of war," he demanded that the nation invest heavily in development of very expensive wind, geothermal, and solar power alternatives. Though most contemporary projections determined that these energies would not be commercially viable until well into the twenty-first century, Carter believed they would speed up the construction and operation of conventional uranium-depleting nuclear plants. Americans would pay prices far higher for their energy than advocates of the breeder estimated necessary. For the immediate future, America would remain dependent on foreign oil.[4]

Response to Carter's program ignored its moral underpinnings and concentrated on its ability to further a group's particular cause. Representatives of the Sierra Club, Friends of the Earth, Environmental Action, and the Natural Resources Defense Council all hailed Carter's proposal for its "fundamental reliance" on conservation and its orientation toward "local initiatives and decentralized energy sources." Those from the Scientists' Institute for Public Information and the Critical Mass Energy Project claimed that the program was simply a backdoor way to establish nuclear energy, including the breeder reactor, as the nation's predominant energy source. Nader's interests concentrated on Carter's failure to provide detailed mechanisms that would prevent energy conglomerates from dominating the new energy plans as they had the old.[5]

Reorganizing the principles upon which to conduct governance created a maelstrom of activity that reoriented many aspects of the federal bureaucracy. Areas previously considered essential now were marginal and vice versa. Nothing could have been a greater boon to science and technology. Rejection of techniques, principles, and areas long established placed a premium on the creation and introduction of their replacements. In the case of power, breeder reactors were out and foreign oil greatly diminished. Coal, conventional reactors, conservation, and "alternative" energy sources gained the federal research dollars as America's well-being and prosperity now depended on the success of researchers in these areas. These hardheaded scientists did not focus on the morality of their work—that was the province of Carter's idealistic policy makers—but concentrated on translating their efforts into practical results. In that sense, Carter's scientific effort was innovative and restorative in that it tried to fashion products and practices not as yet established in areas long explored.

Congress was less charitable with Carter and his fundamental reorganization. On its most basic level, Congress found itself at odds with Carter's insistence that Christian morality govern American policy concerns. While some saw that approach as naïve, shortsighted, or the like, others considered it antagonistic to participatory democracy. It replaced analysis of what was in the commonwealth's interest with broad, religious-based precepts, and Congress sponsored several studies to point out the economic shortcomings of Carter's proposals for the American citizenry. Still others found Carter's presidential campaign as an outsider, pledging to remove the bloat of special interest from government, as hubris, a personal affront to their hard work akin to slander. For these and other reasons, Congress was unwilling to grant Carter the latitude it granted other presidents.

Carter first unveiled his energy platform in March 1977. Before the end of summer, Congress had adamantly recorded its dissent. Senate committees refused to endorse new oil and gas taxes or natural gas price controls. The House repeatedly authorized moneys for the breeder reactor—the Senate assented—even as Carter vigorously vetoed the bills. Congress did, however, approve creation of a Department of Energy to rationalize energy policy, and Carter moved all but the nuclear and environmental energy programs, which his public interest advisors thought needed to retain autonomy to act as effective counterweights, under its umbrella.

To a large degree, Carter's campaign against the breeder reactor held symbolic importance. It signaled his desire to create a moral universe, a universe based on Christian precepts, and a universe in which America functioned as an active and willing partner. His rejection of a profitable, simple means of energy production for higher prices, dependence, and shortages would grant the nation the status of moral arbiter, a shining beacon from which all other nations would gain insight. If the world's foremost superpower cast out its demons of greed, vanity, and hubris, then all other nations could follow suit.

Yet his choices quickly went afoul of special interests. As Carter's Energy Department began to systematize research and development, environmentalists complained that solar energy received too little emphasis and oil-based fuels and nuclear power much too much. They contended that environmentalism was sacrificed for quicker fixes. Nuclear interests also railed against Carter's programs. They disliked paying higher prices for new uranium when they could breed their own and chafed at the restriction that their technology or product could not be sold abroad. Finally, they claimed that Carter dragged his feet on licensing new conventional reactors and thus hurt their bottom line further. Government scientists complained of the redirection of the national laboratories. Traditionally places of weapons and nuclear reactor research, these entities had had great latitude in setting their particular agendas. Under Carter, laboratories became more accountable, their duties neatly defined. Their tasks included research and development of alternative, especially synthetic, fuels. National laboratory scientists chafed at being reassigned and closely monitored, while other segments of society feared that the laboratories' new developmental mission would place them either in competition or cahoots with private enterprise.

Christian morality played a fundamental role both in the implementation of Carter's energy scheme as well as in its formulation. For example, Carter's demand that Christian ethics guide American policy contributed to oil shortages and rising prices. Carter set human rights as the principle around which to conduct foreign policy. The shah of Iran, pressured by his partner, the United States, had cut back sharply his oppressive regime to placate the super-power. Fundamentalist elements unified anti-shah opposition and subsequent attempts to restore the shah's authority failed. His abdication on January 16, 1979, became a foregone conclusion. A Muslim fundamentalist regime openly hostile to the United States replaced him. Not only did the United States lose access to Iran's oil, but also other Middle Eastern principalities raised prices to profit from the shah's demise and refused to kowtow to the United States.

The shah's downfall promised to make nuclear power an even more important part of American energy policy, but a scant two months later that dream fell through. On March 28, 1979, the conventional nuclear reactor at Three Mile Island near Harrisburg, Pennsylvania, began to overheat dangerously. As the core heated up, scientists and engineers labored to figure out how to cool the core down. Failure portended a nuclear disaster, anything from a nuclear explosion to a release of radioactivity to contamination of the Susquehanna River and the Chesapeake Bay watershed.

The accident actually released little radiation. But the inability of technicians on-site to diagnose the problem or agree on means to resolve it proved catastrophic for the nuclear power industry's prospects. Indeed, Three Mile Island was exactly the sort of accident that the nuclear power industry and most scientists and engineers pooh-poohed as impossible, hype propagated by nuclear opponents. A significant portion of American society had viewed reactors as dangerous before Three Mile Island. Few thought them worth the risk thereafter. New procedures and precautions were issued for reactors already up and running. But the chances of any new ones being brought on line shrank almost to zero. (See fig. 8.)

Carter recognized the implications of Three Mile Island and immediately decontrolled natural gas and oil prices in hope of spurring new exploration and enforcing conservation. He also issued executive orders mandating tough energy efficiency standards in federal buildings and with federal contracts. Contemporary polls demonstrated that Americans were far less willing to sacrifice than the president demanded. Sixty-eight percent deemed the energy dislocations a simple hoax to get Americans to pay more for petroleum.

http://phil.cdc.gov/phil/details.asp

Fig. 8

On March 28, 1979, a minor malfunction at the Three Mile Island nuclear power plant in Pennsylvania escalated into a partial core meltdown, resulting in the release of radiation. Before the crisis was brought under control, the uncertainty caused significant stress for officials and a tense public. Although the accident caused no deaths or substantive risk of injury to either workers or nearby residents, it reduced public support for nuclear power, especially since the Three Mile Island accident happened to occur just as Hollywood released the film *The China Syndrome*, with a plot centered around a nuclear near-disaster. The episode led to significant changes in nuclear plant operating procedures and tightened federal regulation that had a dramatic impact on the nuclear industry.

Despite lack of consumer concern, Carter pushed ahead with his program, increasing investment into waste fuel recapture, synthetic fuels, biomass conversion, and geothermal energy. He even hailed the coming of a new epoch in human history, the solar age, arguing that "no foreign cartel can set the price of sun power; no one can embargo it."[6]

Even as Carter supplemented nuclear power efforts with other energy initiatives, he faced the problem of what to do with spent nuclear fuel. This highly radioactive waste had no purpose; it could not be converted to bombs or used further to power reactors. It had been kept at the reactor sites, places hardly secure from contamination or mischief. Storage costs were in excess of $600 million per year. Carter polled his agencies to formulate a permanent solution but when he received conflicting advice he established a committee to settle the matter. It too offered no conclusion. To be sure, the various scientists and engineers recognized the need for a radioactive waste dump. But they could not agree on how to store the material for millennia without risking water and soil contamination or how to prevent sabotage. Should it be stored above ground or beneath it and in what substance? What material would prevent deterioration for the lifetime of the waste? Some political leaders demanded a cessation of nuclear reactor use until these problems were resolved. Others looked ahead to the political problem of locating the waste dump somewhere. No one wanted the dump in their backyard and states retained formidable power to resist federal advances.

Carter hedged his bets in early 1980. He announced that a waste dump would be created and that the waste would be buried in containment vessels. He then announced eleven different sites, each selected for their geological permanence—earthquakes would be catastrophic—and consistency in seven different states. He also formed a federal council to attempt to negotiate with states "as partners in the program" to accept the wastes.

SCIENCE AND SOCIAL SCIENCE TO THE RESCUE

Carter's energy initiatives and problems suggested the circumstances around which scientific and technical matters seemed appropriate in furtherance of Christian morals. So too did Carter's idea of an Institute for Technological Cooperation, which was an extension of the premises of Kennedy's World Food and Nutrition Study. Here the goal was not to create new technologies or sciences but rather to use them to respond to the developing world's unique situations. Developing nations would bring their practical problems—to the United States and there the government would pay—in effect do it in-house or by contract with leading universities—to have the research performed and devices created.

Carter's science advisor, Frank Press, strongly favored this program. Carter had gravitated to Press as his science advisor in early 1977 primarily because Press held expertise in the areas Carter had hoped to move—Press was a seismologist who had worked to detect underground nuclear explosions and understood various types of petroleum exploration and production—and he shared the president's passion to employ science to solve real-world problems. Almost immediately, he joined Carter in extending the concept of planning. To both Press and Carter, planning appeared probable because it seemed objective; it was simply measurement, analysis, and application. According to Press, the new-for-science-advising process mimicked what had long been done within the defense establishment and began with a hypothetical set of questions. These were "what if" or "how would" questions, essentially scenarios logically developed by extending current concerns to several of their possible conclusions. In each and every instance, Press's policy apparatus could weigh and formulate policy to be implemented in case any of these scenarios seemed to be coming to fruition.

This type of planning and policy work was meant to be flexible, dynamic. The vast majority of the scenarios plotted would never occur. But in all cases, fundamental principles and goals remained constant—planning assured that the moral ground could be staked out no matter what the decision. Only the methods to achieve them changed; these methods depended upon any number of situations, variables, or unforeseen circumstances. What this type of planning/policy formulation did was try to anticipate every conceivable situation because its practitioners believed that anticipation would enable them to apply social and other sciences in a rational, consistent manner. It was a style of analysis directly tied to its practitioners' imagination. Put more bluntly, Press and Carter hoped scientifically to predict variable outcomes and a variety of outcomes in every case, thereby achieving relative moral certainty whenever a charted scenario actually occurred.

Among the thorniest questions confronted by Press's analysts was how the United States should respond to a persistent imbalance in foreign trade. To be sure, reliance on foreign oil was a significant factor and Carter (as we have seen) authorized considerable research funding targeted toward creating alternative fuel sources to reduce American dependence. A huge import imbalance in automobiles and also consumer electronics as well as other goods also drew attention. What seemed particularly perplexing was that the vast majority of this deficit went to Japan and Germany, two nations defeated during World War II.

Japan seemed most surprising. Prior to World War II, the nation had been known for producing inferior goods and various explanations surfaced for its postwar renaissance. Some pointed to an alliance of government, labor, and industry, which they claimed worked to set priorities—to plan a foreign economic policy—and to assist the nation in gaining industrial preeminence. In this framework, Japanese industry offered its workers lifetime employment, which reduced turnover and lessened class tensions. Corporations themselves emphasized profits less than their American counterparts, choosing instead to focus on value-added exports. With government aid, they adopted sophisticated manufacturing and assembly techniques and stressed knowledge-intensive and energy-economizing products with high markups. In these pollution-free, state-of-the-art sites, color televisions, calculators, and videocassette recorders rolled off assembly lines directly for the American market.

Congress initially confronted these "unfair" Japanese practices with tariffs or import quotas designed to protect American industries. But Press turned to social science. Rather than restrict Japanese imports, he sought to analyze that country's advantages and to work to reorient American practices to outpace them. Two major areas immediately presented themselves: how could the federal government assist industrial innovation and how could it enhance American productivity?

The first of these questions seemed preeminently a question of science. How the government could spur innovation immediately led to a consideration of basic research funding. Carter claimed that real basic research funding had dropped precipitously since 1970 and had put substantial increases in his budgets. As significant, he instituted zero-base budgeting into the process. Using a formulation created by the social science/management firm of TRW—its CEO, Simon Ramo, had been a Ford science advisor—the zero-based practice required justification not simply for any budget increase but also for the base—the previous appropriation—upon which the increase would be added. Designed to achieve maximum impact with limited funds, it discontinued or rearranged practices that could not be defended or sustained in favor of placing those funds in areas that seemed more promising. Carter's attack on pork barrel science—funding distributed because of political connections rather than national needs—followed from the same vein. He deemed such spending not only wasteful and inefficient but also immoral, akin to stealing.

Increasingly, moreover, research and development (R&D) became synonymous with basic research in the Carter lexicon. Perhaps as an outgrowth of the thinking required in zero-based budgeting, R&D accepted the assumption that research resulted in new products and processes but accentuated the immediacy of the connection; in an R&D world, basic research needed to be targeted toward something that would yield almost immediate payoffs.

This sort of analysis brought attention directly to the national laboratories and universities. Through contracts, grants, and other means, scientists conducted research at those sites. Yet these discoveries and the inventions based upon those discoveries remained the property of the funding entity—the federal government. Government sold licenses to use these inventions. That practice was democratic in that it was open to all but in practice it favored those with enough money to speculate. As important, less than 4 percent of federally funded inventions were ever licensed. The rest remained undeveloped. Carter proposed that institutions at which science was done with federal funds could keep their intellectual property if they used private money to translate that science into practice. Specifically, they retained patent rights simply by using private funds to develop and market a product. In this way, new products and practices would not languish but reach consumers expeditiously and regularly; innovation would flourish, revitalizing American industries, and all Americans would reap the material benefits.

Congress passed this measure in the last days of the Carter administration. Well before the law's passage, however, the president promised that government would foster "a new surge of technological innovation by American industry" and at Press's behest instituted a major domestic policy review to identify innovation bottlenecks. A cabinet-level study, the review committee drew its members from fifteen federal departments and agencies and made provisions to gather material from industry, labor, and public interest groups. Industry argued before the committee that the federal government created numerous disincentives to innovation. Claiming that a series of "stifling constraints" caused it to hesitate or to pass over possible innovations, industry cited an antiquated tax code, antitrust case law, and inflexible regulatory procedures as primary culprits in America's innovative lag.[7]

Press's social scientists seconded the industrialists' sentiments, but, in an attempt to transfer the success of the Japanese model to American culture, they also focused on developing a more prominent role for government in an innovation/productivity partnership. At Press's recommendation, Carter proposed creation of jointly funded government-industry technology centers to investigate problems, such as corrosion, that affected many materials and industries. Press also suggested quintupling to $12.5 million per year of an NSF university–small business innovation program. Starting from the assumption that more than half of all industrial innovation came from small businesses, the plan would fund incubators to speed the development of innovation and its introduction into industry.

Each proposal had its critics. Industry argued that the proposals did not go far enough. But of greater concern was the response of environmentalists, liberals, and other public interest groups worried about the prominence of business within American society. These men and women comprised Carter's base. His willingness to consider reducing taxes on corporations to stimulate innovation, relaxing antitrust law, voiding environmental safeguards, and partnering with industry suggested that the outsider/social justice platform he had run on was little more than lies. Wary that the review was merely a "vehicle through which corporations can complain that it is regulations that are impeding innovation" and that the "values of business [are] penetrat[ing] the values of government," critics argued that "public accounting" must shape all innovation initiatives. They defined this accounting as similar to cost-benefit analysis as they demanded "good social indicators fashioned to . . . evaluate the social costs and benefits" of each proposed innovation. Environmentalists in particular blasted the administration and mulled over the possibility of withdrawing their political support.[8]

By late 1979 the general industrial malaise had been reconceptualized not as a failure of innovation per se but of productivity generally. Carter's overtures to China as he sought to open new markets raised the specter of creating a new economic juggernaut. The rate of increase of American productivity diminished by more than half during the previous decade and for the last year or so had become flat. This all occurred as the rate of increase in productivity among Japanese workers grew at a sustained 6 percent per year.

Several explanations were offered for the discrepancy. Some economists targeted inflation but others examined American society to determine the culprit. Many identified a sociopsychological cause, loss of a work ethic among Americans, and attributed this catastrophic change to a firmly positioned social safety net as well as the success of unions. Together these social factors contributed to easing life, raising expectations, and empowering workers. Labor no longer had potent incentives to get ahead at all costs. Still others pointed to a new hedonism—this was the "Me Decade"—and argued in a more anthropological vein. When favored with alternatives, including the expansion of leisure, Anglo-Saxon attitudes gravitated to a less pressurized, less competitive existence. A decline in the American standard of living would be an inevitable consequence of this sort of thinking, they contended.

Initially Congress dismissed any social science explanation and even social science itself. The House Budget Committee early in 1980 made the strongest statement. It suggested that the government stop funding social science research or studies and devote that money to "basic research and science." Those two areas were "fundamental to increased industrial productivity and economic well-being."

Sociologists took the lead in rebutting the contention that social science research was inherently less important than research that could be converted into consumer products. In fact, they asserted that social science held the key to resolving productivity problems. The House Committee on Science and Technology was instrumental in resurrecting social science research. It held several inquiries into the state of American productivity. A sociologist studying Japanese industry and business for the past decade, University of Michigan's Robert E. Cole offered especially compelling testimony about how Japanese managers differed from their American counterparts and claimed that this discrepancy manifested itself in huge productivity differentials. He made his case simply. In the United States, "a strong tendency" exists "to underestimate the potential of harnessing worker cooperation to raise productivity and to improve quality." As a consequence, he concluded, "we underestimate the contribution to be made by the social sciences." Part of the problem, Cole argued, was the difficulty in measuring "human effort or commitment." Government officials and industrial managers would rather concentrate on easily quantified variables such as research dollars per unit output.

That practice was wrongheaded. The productivity problem lay with managers and officials, who thought of employee-employer relations purely in adversarial terms. They could not conceive of co-operation and motivation as positive elements. They found it easier to "invest millions to make machines idiot proof" than to work with laborers. The Japanese were radically different. There they "encourag[ed] workers to approach their jobs with integrity and commitment and are getting excellent results." Japanese managers "recognize social science and organizational research as relevant to their needs."[9]

There was a remarkable irony in all this, Cole continued. Much of the social science used in Japan came from scholars in the United States. Quality circles rested at the heart of this cooperative/motivational effort. Japanese workers got together on a regular basis to redesign aspects of their work. Each brought suggestions of how to reduce defects, lower costs, and increase productivity. Japanese automotive manufacturers received an average of nine suggestions per year per employee and adopted more than 80 percent. American firms got considerably less than one suggestion per year per employee and accepted less than 25 percent. The consequence was that Japanese workers felt valued and worked hard for the company.

As social science was demonstrating its vitality, the engineering professions became another claimant for federal attention. Maintaining that the NSF was biased toward basic research and therefore stultifying applied research and engineering, they proposed an "equal rights formula" to level the playing field; the engineers called for Congress to establish a National Technology Foundation. This entity would require no new funding. It would simply "recapture" money from the NSF and from the Commerce Department.[10]

Not surprisingly, the NSF's governing board vehemently disagreed with the engineers' proposal. If Congress would have broadened the NSF's mandate and increased funding, they would have created a directorate of engineering and a directorate of social science within the NSF. Engineers thought this suggestion was too little, too late, but university scientists attacked it from the other end. They worried that Congress would welcome increased NSF emphasis on applied research and use the occasion to decrease basic research support. Persons from several different persuasions wondered if this was not an excellent opportunity to examine the NSF's charter and to decide what revisions were necessary.

GETTING GOVERNMENT OFF SCIENCE'S BACK?

The national elections of 1980 took center stage as Congress mulled over the situation. Critics contended that Carter had been overwhelmed by events. In a poignant moment, he blamed his situation on a nation that had lost its confidence. A malaise swept a country racked by assassination, the collapse of South Vietnam, Watergate, hyperinflation, and an energy crisis. This profound sense of disease (and a potent sense of opportunism) led Senator Edward Kennedy to challenge his party's sitting president. The New Deal constituencies upon which Democratic power had rested for nearly half a century felt betrayed by Carter's unconventional approach to social problems and his apparent lack of success. Kennedy attracted many of these East Coast, inner-city Democrats. California's Governor Jerry Brown, memorialized in the comic strip *Doonesbury* as Governor Moonbeam, joined the fray and aimed at the New Left— young, Western, New Age environmentalists.

Fractured by geographic and ideological lines, the party had no chance. Of course, Carter had helped usher in the era of hyphenated Americans when in 1976 he declared himself both a Southerner and an American. Replacing the idea of America as a people with the idea of the nation as an agglomeration of subpopulations, each with its own uniqueness, virtually guaranteed that the eventual nominee would be defeated. Party discipline could only be a pipe dream. A Carter aide said it simply. "We have a fragmented, Balkanized society" with each group "interested in only one domestic program"—its own.[11]

The Republicans offered Ronald Reagan as an alternative. Running an optimistic campaign to restore American preeminence by "getting government off our backs," Reagan appealed to an America before it had its current problems, an America of might and right, and fingered government itself as the culprit. Reagan's program, adherents agreed, would recreate "morning in America," the time when the nation was entering its ascendancy and poised to lead the civilized world.

Reagan's campaign against big government terrified many scientists and engineers. The long-standing government-science nexus depended directly on huge amounts of governmental largesse. Reagan's verbiage threatened to end this connection, "unleashing private enterprise" to pay for virtually all research and development. Scientists and engineers working in defense-related areas saw Reagan differently, as instituting a program of modernizing America's nuclear and other weapons by providing federal dollars to stimulate those areas of inquiry.[12]

Not surprisingly, Reagan's scientific advisors during the campaign came predominantly from defense-related and other industries. Many had had experience in previous Republican administrations. Whether acknowledged or not, science advisors had become identified with political parties. This tension often came to a head whenever scientists held meetings. The questions often asked were not about whether one party or the other was dangerous but rather about the relationship between science and government. Democratic supporters pushed science and technology as a remedy for social and other ills and forcefully moved to apply them into those areas. Republican scientists favored military-involved work. To this latter group, it remained the province of industry to perfect research into tangible products or processes.

Reagan's election did little to calm scientific fears. Month after month passed without appointment of a White House science advisor, an ominous sign for those devoted to maintaining the federal science connection. Rumors circulated that Simon Ramo or Arthur Bueche, GE's senior vice president for Corporate Technology, would be tapped for the job. Neither had the research science experience or academic connections that had marked earlier advisors. Scientists fretted about Reagan campaign promises to abolish the Departments of Education and of Energy. They suspected that the EPA would be eliminated and worried that a massive arms build up would destabilize USSR–USA relations and lead to nuclear war. Within a month of taking office, Reagan confirmed some of the scientists' fears. He dismantled Carter's energy program by ending mandatory energy conservation in federal buildings, withdrawing federal energy-saving standards on new appliances, and renewing breeder reactor research.

Reining in federal spending constituted Reagan's first sustained initiative. Putting that task in the hands of David Stockman, formerly a Michigan congressman and now director of the Office of Management and Budget, the Reagan administration announced across-the-board cuts for the next fiscal year. Its proposals for science-related items concerned scientists. Reagan planned a dramatic increase in military R&D—significantly heftier than that recommended by Carter—but lowered Carter's civilian science proposals by more than a billion dollars. Still, the Reagan civilian science effort increased proposed spending by 21 percent over Carter's final implemented budget.

How the civilian cuts were to be distributed further angered scientists. Stockman wanted to eliminate or drastically curtail programs to upgrade university laboratory equipment, to foster international cooperation, and to train women and minorities. The NSF's social, behavioral, and economic science commitment was scheduled for a 50 percent reduction. Scientific experiments planned for the new NASA space shuttle suffered considerable hits, but the shuttle itself received more generous funding than Carter would have provided, a signal to many that the administration was planning to use the vehicle to militarize space.

Scientists responded by deeming these cuts "vindictive, arrogant and ignorant." That they would object so strenuously to Reagan's targets reflected the vast change in federal science since the mid-1960s. Then physical scientists objected to opening federal programs to social science research. Now an attack on social science research appeared as an attack on science generally. Federal planning to rid America of social problems had become so ingrained as policy that to suggest that it be reduced or even examined seemed a rejection of all science. Members of the National Science Board—Carter-appointed scientists that nominally provided NSF oversight—threatened to challenge legally the administration's handling of the NSF because they had not been given the chance to participate in the new NSF budget. Maintaining that Stockman's budget was the consequence of "star-chamber procedures," *Science* offered a more damning assessment. "For the first time," noted William D. Carey, the AAAS's executive officer, "summary judgment has been passed on the legitimacy of particular fields of scientific inquiry without due process. The social and economic sciences have been scored as flunking tests of need and worth on the scale of government fiscal values."

As troubling was "the implicit judgment that science has nothing useful to say about contemporary dilemmas and issues." According to Carey, the charge against the social sciences "is that they are esoteric." That was nonsense, he suggested, because many of the nation's challenges were economic and social. To Carey, the matter was clear. "Isolating the social and economic sciences [as unworthy] means inflicting damage on integrity of" all scientific research.[13]

Neither Carey nor other politically attuned scientists were mollified when in mid-May word leaked out of Reagan's selection of a science advisor. George A. Keyworth, a nuclear physicist working at Los Alamos government laboratory and a favorite of Edward Teller, the father of the hydrogen bomb, appeared to them as "not a member of the scientific establishment" and therefore unable to "provide any channel between the national scientific community and the White House."[14]

That Keyworth lacked "obvious credentials" appalled them but his public statements made them cringe. He confirmed that economic recovery was the most important challenge facing America, even at the expense of constraining federal R&D support. He further defined himself as a "team player" and his White House post as "advisor to the president, not a lobbyist for science." He promised only to consult with "the top scientists . . . committed to the goals of this office." Arguing that "it is no longer within our economic capability, nor perhaps even desirable, to aspire to primacy across the spectrum of scientific disciplines," Keyworth called on his fellow scientists to help him "identify those disciplinary areas where vitality is required to support industrial, military [technologies] . . . as well as those with particular scientific promise . . . measured in terms of probability of major breakthroughs." For his part, Keyworth saw molecular biology, genetics, weak interaction physics, agricultural research, computer science, and mathematics as those "disciplines [that] should receive the greatest support . . . during time of limited budget." And he called on his colleagues to think about "the real impact" of budget cuts in the social sciences, not just that they would be achieved disproportionately. Keyworth suspected the impact to be minimal. His desire to have the OMB end its "cost and time accounting issues," which "requires scientists to account for every moment of their time" was perhaps the only item that would find scientific favor.[15]

In the few months after Keyworth's appointment, the Reagan administration announced several new policies. It called, for example, for removal of impediments created in wake of the Three Mile Island fiasco to licensing nuclear power plants. Deregulation of domestic oil prices followed from the assumption that energy policy must reflect market forces, not federal rules. Conservation, it argued, would be the natural by-product of deregulation as would stimulation of private investment into solar, wind, and synthetic fuel technologies. Government programs would be unnecessary. Reagan also relaxed federal export regulations for nuclear and especially breeder technologies, maintaining that political and diplomatic initiatives would prove more effective in ensuring global stability while enabling American manufacturers to profit from selling their technologies overseas. The administration tried to terminate an NSF technology transfer program enacted in the last days of the Carter administration as something outside the federal government's proper province and announced a $180 billion program to "strengthen and modernize the strategic triad of land-based missiles, sea-based missiles, and bombers" to "encourage accommodation and prudent behavior on the part of the Soviet Union."[16]

Resistance to these and other Reagan science initiatives came from an unusual source. Frank Press, formerly Carter's science advisor, assumed leadership of the National Academy of Sciences and quickly positioned it to confront the administration. Among Press's first acts was to reject the time-honored tradition of only replying to government requests. Instead, he had the academy identify and study "critical scientific issues." Similar to his approach as science advisor where he examined scenarios to anticipate situations before they occurred, Press used the academy to produce position papers on possible future consequences of present-day science policies. This thinly disguised critique of Reagan science was followed up by a more direct affront. Press called an "unprecedented gathering" of American science leaders to "control the damage" done by the administration's budget-prioritizing policies. Using "rational and balanced" judgments, the conclave would identify "alternatives" to Reagan's initiatives, thereby insuring "wise allocation" of meager resources.

About one hundred scientists and engineers attended the two-day October meeting. After offering "horror stories of canceled experiments, of discouraged graduate students drifting off, of infighting among colleagues," the group claimed that Reagan policies were shortsighted, causing "irreversible damage," likely to doom America. Allen Bromley, a Yale University physicist and president of the AAAS, contended that "'guerrilla warfare' had begun to break out among scientists fighting for limited funds." America's socio-economic malaise necessitated "more, rather than less" science funding and the assembled demanded "a much strengthened mechanism" for scientists and engineers "to advise the government on resource allocation and the impacts of various budget strategies."

The conclave's demands marked a confluence of several different assumptions. First, it suggested that science and scientists were objective and therefore they would use their unique methodology to "scientifically" assign government resources. But this technocratic vision, which included a legislative element since allocation was part of the plan, was in apposition to Press's stated desire for Congress to overrule Reagan science initiatives. The *Wall Street Journal* recognized the inherent dichotomy. "Scientists wanted to influence federal budgeters," the journal claimed, "without sounding like just another special interest group whining about being pushed away from the taxpayers' trough." The group openly worried, according to the paper, that they should not "make the manifesto sound 'too self-serving.'"

Keyworth and other high-level Reagan appointees attended the meeting and staunchly defended the administration's scientific policy. Keyworth accused those gathered as "lacking in realism," victims of "panic" and "paranoia." In tough budget times, choices needed to be made to support only those sciences and technologies with "maximum promise" and "clear relevancy." American science, he continued, was mistakenly "equating dollars spent with quality." This led to "tolerance for mediocrity and less stress on excellence." The associate director of the OMB put the matter in the context of the federal budget crisis generally. Science was "absolutely flourishing" compared to other programs but the administration did not believe "everything labeled basic research had intrinsic merit." Choices had to be made.[17]

The heated, contested meeting suggested that for the duration of the Reagan administration, scientists and Reaganites would be at odds. Despite considerable rhetoric and posturing otherwise, that proved not to be the case. Within a couple of months, Keyworth established a science board to advise the president and made sure it represented science broadly, not just a sole perspective. He also recognized the administration's social science budget cuts as too drastic and vowed to restore funding during the next fiscal year. The Department of Defense offered several new programs to increase the number of graduate fellowships in science and initiated new laboratory refurbishment grants. The administration reversed its decision to decrease aeronautics funding. Reagan himself championed the space shuttle and called for a more permanent presence in space.

Scientists continued to grouse. Some argued that the nation would lose the race to the stars. Others chafed at the stiff new rules imposed by the national security–conscious Reagan on classified research as well as the expansion of the definition of that research, while still others worried as the administration cut back on USSR–USA joint science ventures. Environmentalists continued to attack the administration for its proposed EPA regulations and consumer advocates saw the FDA as less responsive to their concerns. The administration drew fire for its efforts to spin off its successful Landsat program, which used satellites to check on crops, search for mineral deposits, and measure atmospheric pollution.

GETTING SCIENCE OFF GOVERNMENT'S BACK

Yet scientists and the administration found considerable common ground. In early 1982 NAS president Press introduced a novel plan to guarantee science's federal future. His campaign for long-term federal basic science support included a stipulation that government must increase funding of basic research at a minimum of 2 percent above the rate of inflation each year. New research facilities would come from a separate budget. In return for this guarantee, scientists would work with government to identify less productive research institutions and to help move funding from these less productive schools to more productive ones. Industry would figure in this scheme, providing a minimum of $50 million yearly to support university R&D. Together industry, universities, and the federal government would maintain graduate science education support.

Press understood that what he proposed was a science entitlement, setting science up outside the normal budgeting process. But "we are raising a new generation of Americans that is scientifically and technologically illiterate," he complained. His plan derived its justification from the fact that a "good fraction of the US gross national product is due to new knowledge." Therefore, basic research "is not science for its own sake." "There are very important national security matters involved." And the science budget must always get larger; it can never remain the same. Science "becomes more expensive with time because, as science progresses, the problems it addresses become more complicated. This is a built-in inflation factor," he concluded.

Press's plan was path breaking but perhaps its truly revolutionary aspect was that it was formulated in conjunction with Keyworth. Together they framed an agenda that addressed the area upon which Reaganites and scientists agreed: the centrality of basic research to the nation's weal. Keyworth took the case directly to industry. He maintained that the government's "massive" basic science investment was "good for pure knowledge, but not so good for industrial needs." To get this work to the place where it could be developed, businessmen needed to "knock down some of those walls between industry and universities." But industry also needed to get involved in basic research. Corporations must learn "that concentration on short-term goals and neglect of longer-term research has left many technology-dependent industries poorly positioned for future competition with foreign industries." Microelectronics seemed the next technological frontier and Keyworth cited industry's increase of R&D expenditures (16 percent in 1981 and an estimated 17 percent in 1982) as a favorable indicator that the nation was starting to get the science investment message.[18]

But even before the Press/Keyworth initiative, *Science*'s William Carey had proclaimed the "worst is over." He now believed that the federal government would not recant on its science commitment and the public now recognized science as something worthy of support. Science, he proclaimed, was quite healthy in comparison to other parts of the economy. Yet he too understood Keyworth's contention that in science, as in arboriculture, "occasional pruning . . . can promote, rather than retard its health," and he admitted that the seemingly unassailable "postwar construct of the R&D enterprise is worth thinking about." It could be, he predicted, that the recent "economic stress" on science could cause universities, industry, and the federal government to reconnect in new, greater ways.[19]

The basic science research accommodation did not mark a fundamental shift in Reagan's policies. In fact, it was a natural consequence of Reagan's philosophy of governance. To the Reagan administration, government came from a single precept: its duty was to accomplish only those things that individuals could not. The way that the administration operationalized that principle was by a sort of means testing. In each instance of possible governmental action, it began with the negative, the assumption that government for that purpose was unnecessary. In effect, Reaganites believed in zero-based government; it was the responsibility of the petitioner for government to prove that government would be essential to achieving the particular task *and* that that task was a legitimate responsibility of government.

Reaganites were no less wedded to their philosophy than Carter was to applying Christian morality. But there the similarity ended. Reagan's philosophy was highly rationalist and incorporated modification of policies as the situation dictated. Carter was doctrinal, uncompromising. That they are remembered conversely stems from their style rather than their substance. Reagan continuously espoused his first principle and used draconian sound bites such as "getting government off our backs" and "government that governs least governs best" to press his case. Carter merely wanted to have a "government as good as the American people." The result was a perception with little basis in reality. Carter seemed wishy-washy, while Reagan appeared resolute, a countenance that inspired great loyalty.

This perceptual anomaly also helps account for the fierce antagonism the accommodating Reagan engendered. In this context, every measurable sign that the administration had backed down from its previous position seemed a victory of momentous proportions. To these men and women, the Reagan administration was so inalterably opposed to using government in any constructive manner that the smallest sign of support for a project must be due to lobbying and constituency building, an outcome for which any group through their efforts and the efforts of their friends could be proud. It encouraged fervency among the opposition because only a full-scale onslaught could muster the force necessary to win the day.

Reagan's zero-based government increased the number of stakeholders in government generally. By establishing a forum or contest as the means to adjudicate if government was "appropriate" for a task, Reaganites institutionalized discussion and debate as part of the governmental process. But while debate was encouraged, it was not necessarily democratic in the sense of equal voices or one man/one vote. The rationalist Reaganites favored numbers, statistics, measurement, science, even as Reagan himself frequently descended into anecdote or stories as the "Great Communicator" made his case to the American people. Reagan's stakeholders were leaders, leaders in business, the military, science, law, industry, and the like. It was their arguments and opinions that mattered.

Morning in the Rotunda

Congress flourished within this system. For the first time in decades, the Imperial Presidency was no more. Congress made tangible differences in policy area after policy area. Nearly one hundred Senate and House panels dealt with science and technology matters during the Reagan years. Public hearings served their original purpose, to be forums to discuss the merits of any particular case. Medicine, among the most immediate, personal sciences, attracted much attention as Congress repeatedly tried to specify what areas NIH scientists should study. Despite Congress's new assertiveness, Reagan vetoed congressional measures at a rate virtually indistinguishable from his predecessors. He overwhelmingly accepted the vast majority of congressional policy modifications. Acquiescence had its costs. During the

Reagan administration, the federal budget just about doubled, from just under $600 billion in 1982 to slightly more than $1.1 trillion in 1989.

Part of the federal expansion and a potent demonstration of Congress's new authority stemmed from that legislative body's willingness to earmark significant money for university-based facilities to pursue science and technology. Lobbyist firms began in 1983 to specialize in getting earmarked money to establish new national centers or laboratories at specific public and private colleges and universities. Other schools soon worked with their congressmen and congresswomen to catch up and to guarantee that they received similar perks. The American Association of Universities, the National Academy of Sciences, and the National Association of State Universities and Land-Grant Colleges all decried this pork barrel science, as did numerous scientific and technical organizations, including the American Physical Society, the Council of Scientific Society Presidents, and the AAAS. Each claimed that all funding should be on a peer-review basis. Congressmen from districts receiving relatively little in the way of federal largesse demurred. They argued that the peer-review process remained biased toward those schools that had long been successful with that approach and that earmarks were the only "legal, ethical" way to redress the imbalance. By the mid-1980s, Erich Bloch, NSF director, with less than professional subtlety began to question whether those places successful in the earmarking quest should be entitled to NSF and other peer-reviewed funding. Bloch argued that pork barrel earmarking undercut peer review and thus should eliminate recipients from that sort of open competition.

But this earmarking/peer-review confusion also worked the other way. Congressional studies suggested that money earmarked for agricultural research at the nation's land-grant colleges could be better spent if the process was opened to peer review; private schools and state universities should compete with the traditional recipients of federal agricultural research funds. Still other studies cast doubt on the success of federal laboratories, especially those administered by specific universities. Rather than continue to provide funding to these ineffective creatures and to restrict this funding to the work performed there, some congressmen proposed opening funding to anyone seeking it and relying on peer-reviewed grants to increase governmental efficiency.

Every science initiative Reagan had opposed or adjusted in his first year he accepted and fully funded within the next few years. As early as April 1982, Reagan himself had identified science and technology as core elements to American success, a vision that would persist throughout his presidency. "Science and technology are essential," he maintained, to accomplishing "the goals of this Administration and the needs of the American people for jobs, enhanced national security, increased international competitiveness and better health and quality of life. The continual advancement of both theoretical and applied scientific knowledge is of vital importance to continued human progress and the resolution of the complex problems facing the world in the years ahead." There is, he continued, an "important role of the federal government in supporting our scientific enterprise." But some things "can best be done by the private sector. I believe," Reagan concluded, "that together we will be able to harness science and technology to meet the needs and aspirations of all our people."[20]

Within the Reagan White House, scope mattered. Small projects could be handled by private enterprise. The federal government invested for the nation; it did what private enterprise could not. Hence, the size of Reagan's science initiatives was unprecedented. While it would be incorrect to suggest that Reagan caused the era of Big Science, he did indeed foster it. The greater the vision, the larger the project, the more likely the Reaganites were willing to support it.

The context in which the Reagan administration operated recognized two essential threats. The Soviet Union remained to Reaganites the "evil empire"—a Star Wars allusion—and diffusing or defeating that military menace seemed critical. To Reaganites, the threat from Japan was almost as immediate and more insidious. Its automotive, electronic, and microelectronic successes had promised to render American industries noncompetitive in many world markets. The success of Japanese industries meant that America would have for the foreseeable future a significant trade debt, which contributed to inflation here at home. American military and economic strength had traditionally come from its industrial might. If the Japanese continued to translate modern scientific insight into highly profitable consumer and other products, then America's long-term prospects were dim. It would lack the resources necessary to tackle the Soviets and to achieve the sort of postwar dominance to which its citizens had become accustomed. By early in his first term, Reagan understood that both of these issues revolved around science and technology. For America to maintain its place in the world it was required to meet the very different Japanese and Soviet challenges. And that could only be done by massive expansion of federal science and technology funding.

A drive for secrecy was a concomitant of the Reagan view of the new challenges. To provide data, information, or theory to assist one's rival struck the administration and its business-oriented managers as idiotic. Industrial espionage was the single easiest means to acquire new technologies. To prohibit that sort of operation in the public sector required Reaganites to crack down on any information-sharing possibility. They greatly expanded the definition of what constituted classified research and refused to grant visas to groups hoping to come to America to learn about the newest advances. Scientists rebelled against these strictures. In addition to the practical problem of having to prove the validity of one's work in a university setting, scientists argued that science had no borders. It built on the science of others and to restrict communication was to hamper the scientific enterprise. Reaganites dismissed concerns about the hypothetical scientific enterprise and concentrated instead on the bottom line. They only cared that the Soviets and the Japanese would be hurt by secrecy more than the Americans.

Modernization of defense forces constituted the single largest expenditure of science- or technology-related money during the Reagan years. New weapons systems deferred during the previous two administrations went full bore under Reagan. Each new system incorporated the latest technology: computers, cruise missiles, Global Positioning Satellites, lasers, microelectronics, and other high-tech devices to ensure consistent, dependable function. But

Reagan also focused on what he maintained had been a massive Soviet nuclear buildup in the 1970s. The Soviets had concentrated on offensive delivery systems and Reagan expressed concern the deterrence equation had changed. Previously, Mutually Assured Destruction had kept the two sides at bay. But now Reagan worried that the Soviets were approaching the time where they could obliterate US forces with little risk to the communist nation. To restore balance to MAD, Reagan proposed a sophisticated, expensive, cutting-edge technology-based shield that would prevent Soviet missiles from penetrating American defenses and so keep the nation's nuclear response safe from a surprise attack.

REALLY BIG SCIENCE

Claiming that the massive Soviet buildup of recent years focused on offensive delivery systems, Reagan claimed the nature of deterrence had changed. Now to promote MAD required the United States to protect its ability to respond to a Soviet onslaught. His Strategic Defense Initiative (SDI) was directed to that end. Nicknamed "Star Wars" because of its far-reaching, fantastic, high-tech character, SDI was a shield of computer-controlled satellites, complete with laser devices that would emit particle beams to destroy incoming missiles.

The revolutionary nature, coupled with the considerable technical problems of successfully downing as many as a thousand nuclear tipped and dummy missiles at one time, made SDI controversial. Like the public generally, scientists could be found on both sides of the issue. Some went as far as to suggest that difficulties would make such a system virtually impossible, while others proclaimed the basic technology already existed to construct the shield. Still others postulated that the shield should be impenetrable, but a majority claimed that was impossible and debated how much devastation was permissible when SDI was fully introduced. As significant, they carried their case to Congress. Reagan remained adamant in his support of SDI, so much so that his refusal to abandon the project led to a breakdown in USSR–USA arms reduction negotiations. Several times in the later 1980s, Congress came close to terminating SDI and in every instance cut Reagan's spending requests for the project.

Separate from SDI was Reagan's staunch support for a space station. This platform came with an estimated $14.5 billion price tag and could be used for a number of purposes, including scientific experiments and manufacturing practices in a zero-gravity environment. The romance of space had long captured the American imagination and a station played on patriotic fervor, but since the 1960s NASA had been campaigning that space exploration made sense because it produced numerous terrestrial spin-offs. The agency even published a semimonthly periodical, *NASA's Spinoff*, to trumpet these mostly high-tech, computer-related accomplishments. But never had NASA personnel nor anyone else assessed the validity of the contention. Would the same amount of money spent on other research have yielded greater technological spin-offs? For a number of reasons, Reagan favored the station, regardless of the spin-off question's merits. He also backed a relatively cheap, reusable vehicle, the space shuttle, to build and service the anticipated space station. Originally proposed by the Nixon administration a decade earlier, the shuttle flew for the first time during the Reagan years and could be employed to carry almost anything into space, a fact not wasted on SDI advocates and perhaps Reagan himself. In the shuttle, they saw a dual-use vehicle, suitable for military and civilian use. It could build SDI platforms and place laser satellites into orbit.

Efforts to speed space station construction ended when in 1986 the space shuttle *Challenger* exploded just a few short moments after lift off. The entire crew died in the disaster. The newly revitalized Congress immediately began an investigation of what caused the crash and the shuttle program was put on hold for virtually the remainder of the Reagan administration. (See fig. 9.)

Though not quite as expensive as the space station, Reagan's fervent support of the Superconducting Super Collider marked a Big Science milestone. Proposed by high-energy physicists, whose research money and opportunities in America had dried up, the new project dwarfed any existing particle accelerator and promised to recreate many particles present at the big bang. Its main features were twofold. First, it operated close to absolute zero so that electrical resistance in its huge electromagnets was nearly nonexistent and thus the magnetic force intensified. Second, it speeded particles at right angles along a fifty-two mile underground track to achieve maximum velocity before collision. The project was science for science's sake, but high-energy physicists suggested that the particle accelerator would produce countless spin-offs, unrelated to the high-energy aspects. At first, scientists in other areas were mum about the cost of the project—estimates ranged from $4 to $8 billion—and what that might do to their science budgets. Such was the case in an era of massive federal deficits and even after the 1985 passage of the Gramm-Rudman-Hollings Act, which attempted to get the deficit under control by mandating that increases in spending beyond a target approved by the president and Congress be offset by automatic spending decreases in other areas. The newly assertive Congress itself trimmed Reagan's requests several times. The project set off an orgy of greed among states. Forty-five states applied to have the accelerator located within its borders. Not only would the accelerator create numerous high-paying scientific and technical jobs, but also actual construction of the device would require thousands from the building trades. States provided scientific or infrastructural inducements to get the Reagan administration to locate the new facility within their borders.

The Reagan commitment to reestablish the nation's scientific infrastructure was far more costly than the Super Collider. The administration promised to revamp outdated research and development facilities and create new ones at an estimated $20 billion price tag. Also in his last years, Reagan proposed a doubling of the NSF budget over five years. During the eight years of his administration, the NSF budget had already more than doubled. Basic research received the lion's share of that new funding. To Reaganites, basic research was a sine qua non for government. It was the base from which military and industrial might would stem during successive generations.

The life sciences were not left out in the Reagan Big Science

http://en.wikipedia.org/wiki/Image:Challenger_explosion.jpg

Fig. 9

Although advocates in America's space program had envisioned the shuttle as a "space truck" whose launches would be frequent, inexpensive, and useful, the shuttle program's early successes were mixed with technical complications and practical obstacles. On January 28, 1986, a seemingly routine mission of the shuttle *Challenger* ended after just seventy-three seconds in disaster, after an O-ring seal failure in the solid rocket booster tore the shuttle apart. Public attention to the launch had been especially high, since one of the seven crew members killed was Christa McAuliffe, who had been chosen as the first teacher to enter space. Subsequent technical investigations and expert testimony confirmed that the risks of the shuttle program had been higher than many acknowledged and that poor decision making on NASA's part contributed to the catastrophe. After a two-year delay, NASA restarted the shuttle program, only to have it further set back again by the February 1, 2003, loss of the shuttle *Columbia*.

agenda. A huge project to map the sequence of the three million pairs of genes on the human genome also received Reagan's sanction. Estimated to cost nearly $3 billion and to take twenty years to complete, the Human Genome Project promised to create entire classes of biological and pharmaceutical industries as well as introducing gene therapies and other knowledge-based genetic manipulations.

But even as Reagan planned for the next generation of scientific and technological breakthroughs, his administration tackled the new disease of AIDS. Acquired immunodeficiency syndrome proved a real puzzle. It seemed to burst on the scene and to be invariably fatal. Who it struck was an epidemiological question—the first pocket of infection was noted among San Francisco gay males in 1981—and why and how it afflicted those people also remained elusive until well into his second administration. That it seemed in disproportionate numbers to infect male homosexuals and intravenous drug users led some of Reagan's supporters to proclaim that the disease was the punishment for the wages of sin. Yet by his second administration, government-funded AIDS spending was topping $250 million yearly.

But AIDS was not the only emergency confronted by the administration. Discovery that certain ceramics could exhibit superconductivity at high temperatures set off a scramble to develop technologies based on that knowledge before the Japanese and the Soviets, already at work in the field, could fully exploit it. To that end, Reagan cited a need "to bridge the gap from the laboratory to the marketplace" and announced in mid-1987 a "Superconductivity Initiative" to "help American businesses translate the scientific promise of superconductivity into marketable technologies." Including a series of laws to increase patent protection, relax antitrust regulations, and offer small grants for commercial applications, Reagan's plan provided an initial $150 million to fund four superconductivity centers at national laboratories. This part of the plan built on an earlier executive order requiring national laboratories to make their science more accessible to industry and to work with it to develop commercial products. High-temperature superconductivity "brings us to the threshold of a new age," Reagan asserted, and it was government's responsibility to "herald in that new age with a rush." As if to reinforce the idea that this was an American initiative, the Reagan White House refused to allow foreign scientists and officials to attend its initial major discussion of superconductivity's promise.[21]

Reagan used government to counter Japanese industry in other areas. His "Sematech" program funded joint research by America's leading semiconductor manufacturers because "semiconductor technology will be critical to our economic security in the years ahead." It funded extensive research in robotics, computer-aided design, and computer-integrated manufacturing, all fields in which the Japanese had a substantial record of success. The administration also took on the Japanese in supercomputing. Aware that the island nation already had supercomputers on the market, it unveiled a $1.7 billion, five-year plan to "promote supercomputing architecture, networking technology, and custom hardware."[22]

Each of these new initiatives capitalized on a fear of Japan and remained consistent to the premise of the Reagan administration—government undertakes only those tasks outside the province of individual initiative. Reagan demanded that the nation enhance its competitiveness and singled out the Soviet military and Japanese industry as the nation's two predominant threats. Indeed, in the economic sphere, competitiveness was the Reagan administration watchword. Initially, Reaganesque competitiveness predominantly meant getting government off our backs, reducing red tape, and ending unnecessary regulation. A more favorable patent policy, granting antitrust exemptions to allow competitors to share research expenses and an executive order mandating a regulatory impact analysis for each proposed regulatory action, typified this approach. Patterned after the environmental impact statement of an earlier era, a regulatory impact analysis needed to accompany every new regulation. Each must consider the consequences on society of the regulation. An agency in its regulatory impact analysis must take "into account the condition of the particular industries affected by regulations, the condition of the national economy, and other regulatory actions contemplated for the future" before it could issue a new rule. No regulation could be enacted "unless the potential benefits to society from the regulation outweigh the potential costs to society."[23]

A shift from regulation to research characterized the regulatory policy of the early Reagan years. Rather than check on and force compliance, the Reagan administration concentrated on developing ways and means to eliminate problems. Sometimes federal agencies promulgated new rules. Reducing the permissible amount of lead in gasoline and ending exemptions for new blenders as well as setting the permissible cancer risk of certain pesticides as a hypothetical one case in every ten thousand exposed people versus the earlier one case in every million were two instances of EPA action. In other instances, the administration sponsored research to overcome ambiguity. Under William Ruckelshaus, the EPA worked to "improve methods of assessing risks associated with exposure to toxic substances and ways to decide how to handle the risks." It also initiated a plan to coordinate risk assessment through all federal agencies so that each would share the same toxicological portrait.[24]

Environmentalists balked at Reagan's regulatory approach, arguing that the administration under the guise of reform gutted regulatory control. Just the opposite seemed the case with biotechnology. Citing regulatory discretion, the EPA claimed regulatory power over genetically manipulated substances that served as pesticides, commercial chemicals, and pollution-eating substances. It required persons seeking to introduce any of these biologically artificial microorganisms into the environment to notify the agency at least ninety days prior to that event. The EPA claimed authority to deny any such petitioner. At the heart of this policy were two assumptions. First, extensive evidence existed that introduction of new organisms into the environment sometimes produced massive unexpected changes; a pesticide meant to kill one organism could ruin the entire ecosystem. Second, that genetically manipulated organisms were possibly different than naturally occurring microbes and therefore required more care. No less an authority than James B. Watson, one of the codiscoverers of DNA's structure, thought the EPA's policy ridiculously restrictive. This administration, he maintained, was "just as silly as the others," and he attributed the biotechnology decision to the fact that a woman made it. "They had to put a woman some place," he concluded. "It's lunacy."[25]

WHAT SHOULD GOVERNMENT DO?

Concern over regulatory policy quieted in mid-1983, as the next presidential election cycle began. To counter Reagan's competitiveness campaign and because they understood that the American economy was in trouble, Democrats outlined their own program, which they identified as a national industrial policy. To be manufactured by industry, labor, and government, the national industrial policy would set America's course for the next decades. Reagan railed against this planned economy, arguing that the free market was considerably better for determining which industries should survive and thrive. To pick and choose after a series of negotiations and compromises was to invite disaster. It was a kind of pork barrel policy in which "the American political system" would "pick and choose among individual firms and regions" rather than select industries for their economic merit.[26]

Genesis for the Democratic plan came from the party's analysis of Japan's industrial renaissance, tempered by a respect for the traditional New Deal constituencies. Although specifics varied some between candidates, the Democratic plan focused on modernizing older, smokestack industries rather than abandoning them; providing economic security to labor by guaranteeing employment or funding retraining programs; and establishing a public works program to rebuild America's roads, bridges, and sewers. A superagency, similar to the Japanese MITI and including representatives of industry, labor, and government, would plan, oversee, and coordinate all activities as well as select new, favored areas into which American enterprise would expand.

Walter Mondale's overwhelming defeat in the 1984 presidential election cooled talk of a national industrial policy but Reagan continued to hammer competitiveness. Declaring that the "government's legitimate role is not to dictate detailed plans or solutions to problems for particular industries" he established a commission of leading industrialists to suggest means to enhance the nation's competitiveness. In addition to the sort of palliatives proposed by Reagan years earlier, the panel also recommended a cabinet-level Department of Science and Technology. Like the Democrats, the competitiveness commission acknowledged Japan's MITI. But unlike the other party, their proposed cabinet department coordinated private and public research. Government would undertake basic research, which the American MITI would systematize and make accessible to industry. Industry under the American MITI's watchful eye would then develop that research into marketable products.[27]

Reagan did not move on the cabinet-department proposal, but increasingly business and universities began to implement facets of the administration's competitiveness program. More than one hundred manufacturers joined with the Bureau of Standards in 1986 to make the various companies' individualized, standardized automated metals-manufacturing equipment compatible across the industry. But this was just a small part of the bureau's Industrial Research Associate Program, which regularly brought together bureau personnel and industrial scientists and engineers from a month to a year to pursue research of mutual interest. Industry purchased much of the equipment used in these ventures, nearly $50 million in 1985 alone. The bureau ran a similar program for university scientific and technical employees. Corporate financing of university research nearly tripled from 1980 to 1986. Casimir Skrzytcak, Nynex Corporation's science and technology vice president, put the case simply. Going to universities, he remarked, "is the most cost-effective way of gaining an edge on technology and access to a wide range of thinkers." Often assured of their ability to publish their findings, faculty generally also saw the relationship as beneficial. Not only did it allow them to support more students and therefore expand their operations, but it also enabled university scientists to "play a more constructive role in society" where their efforts could create jobs and otherwise improve the social milieu.[28]

This university-corporation synergy seemed so positive that in late 1986 a joint lobbying effort began. Coretech pushed the federal government to increase tax credits for high-tech research, create a program of incentive and matching grants to modernize university laboratories, and fund graduate research fellowships. Stuart Eizenstat, domestic policy chief in the Carter administration, served as chief lobbyist, and announced that "there is a new recognition" that when a university "is strengthened in its research capacity," corporations "will be benefited too."[29]

In February 1987 Reagan carried the university-industry-government interface to a new level. He announced creation of a comprehensive program of science and technology centers on American university campuses. These centers would be problem-, not discipline-oriented, structured around long-term subjects of interest, such as "robotics for automated manufacturing and microelectronics, new material processes, and biotechnology." But in keeping with the administration's philosophy of government, they would pursue only basic research, albeit on issues "relevant to practical university problems." These centers were organized radically differently than university departments; their problem orientation made it mandatory that there be multidisciplinary research teams. Tearing down disciplinary and departmental walls and boundaries, the centers would "change the very fabric of our research universities"; the centers were "not a program" but "an incipient revolution." In addition to multidisciplinary research teams and problem-oriented missions, the centers included considerable industry participation and accentuated graduate education.[30]

Reagan competitiveness extended to the national laboratories and federal agencies. Scientists in the labs were put in contact with industry and encouraged "to patent, license, and commercialize their research," while agencies inaugurated "royalty-sharing plans" with their researchers. Reagan also promised to "recruit science entrepreneurs to act as conduits between the laboratories and business, venture capitalists, and universities" and to compel the defense and space programs "to spin off technology to industry and to do it even faster than they have."[31]

Competitiveness swept Washington. Congress established a Caucus on Competitiveness. More than one hundred members from both parties claimed affiliation. A Council on Competitiveness was formed by John Young, president of Hewlett-Packard, which included not only the president of Massachusetts Institute of Technology but also the president of the AFL-CIO. TRW, Ramo's firm and several labor unions joined to establish a congressional Economic Leadership Institute to press competitiveness measures. Former senator Edmund Muskie's Center on National Policy, the Senate Democrats' Democratic Working Group on Competitiveness, and the Northeast-Midwest Congressional Coalition, which included House members from the Rust Belt, all made competitiveness their focus.

Congress took the competitiveness issue a step further when in 1988 it renamed the National Bureau of Standards the National Institute of Standards and Technology and gave it new powers to enter into joint research programs with various industries. It also gathered the various programs under way in the Commerce Department, a prime site for competitiveness activities, and placed them under the new Technology Administration, which was headed by a newly created position, the undersecretary for technology. In this way, Congress sought to guarantee coordination between the public and the private sectors as Americans retooled their industrial machine.

The federal government even established a national award to recognize success among American businesses in the competitiveness contest. Named after Malcolm Baldrige, Reagan's secretary of commerce until killed in a rodeo accident, the Baldrige National Quality Award celebrated management as the key competitiveness ingredient necessary "to compete effectively in the global marketplace." The award reified in government the social science concepts articulated by Michigan's Robert E. Cole in his testimony to Congress nearly a decade earlier. Cole had studied Japanese business and identified their management techniques, which valued workers and their insight, as explanations for that country's prodigious and sudden dominance in automotive and electronics manufacturing. The essence of his analysis had become popular in business colleges and schools of management throughout the nation and had been adopted by Ford, General Motors, and other American industrial giants. Congress's proclamation establishing the award echoed Cole's theories. "Foreign competition" has "challenged strongly" America's leadership "in product and process quality." But through "strategic planning for quality," "quality improvement programs," and "a commitment to excellence," American industry could reemerge preeminent. American industry needed to instill in the factory "worker involvement in quality" and had to become "management-led and customer-oriented." Only in those ways would productivity and hence competitiveness increase. Congress authorized the national quality award because a similar program begun in the mid-1950s had spurred Japanese industry to reconceptualize itself. The Baldrige would "stimulate American companies to improve quality and productivity for the pride of recognition" and thus provide "an example to others" throughout the nation.[32]

States quickly jumped on the competitiveness bandwagon, although not always through inspiring new management techniques. Attracted by the possibility of a significant boost in federal funding and under pressure by the citizenry to do something to improve state economies, governors and legislatures in state after state redirected public universities to participate in the competitiveness sweepstakes. They built research parks at state institutions and tried to stimulate high-tech enterprises around universities—there the model was Route 128 near MIT in Massachusetts. A more explicit attempt to expropriate the Japanese experience included the fostering of technopolis, which entailed seeding through tax incentives and outright grants the area between a major city and a state research university some miles away so that an integrated high-tech corridor would grow between these two symbiotic social organisms.

In almost all of these cases, the devil was in the details. Rarely did groups mean the very same thing. Democratic interests tended to see competitiveness as a far more extensive project and a far more protectionist one as they more often sought to shore up the status quo. Despite underlying disagreements and inconsistencies, competitiveness became a major focus and a watchword of the 1988 presidential election.

After Reagan?

Among certain groups of scientists, Reagan's record also was a campaign issue. To be sure, virtually no scientist complained about the

NSF budget nearly doubling during the Reagan administration. Nor did they protest a similar rate of increase in nonmilitary basic research funding. The National Institutes of Health kept pace. Its budget also nearly doubled during Reagan's term and AIDS research funding approached nearly $2 billion for 1989. The military basic research budget stayed roughly the same but money for military development nearly tripled during the Reagan years. Nor did scientists and engineers mind that their share of the work force in private enterprise had jumped from 2.6 percent in 1980 to 3.6 percent in 1988, a 62 percent increase. What gained their attention was how Reagan set science policy. To his critics, Reagan relied on the buddy system—his personal acquaintances—rather than seek out "independent advice on technical questions."[33]

Almost invariably the scientific establishment voiced this strain of criticism. Wiesner, Press, and the NSB led the charge. Each saw a need to resurrect in the next administration an independent science office in the White House where consensus was to be reached and policy formulated. Such an office never existed in any previous administration, of course, but that failed to dissuade these former science advisors. Press and Robert Rosenweig, president of the Association of American Universities, took particular exception to the "Big Science" trend. They feared that the Superconducting Super Collider, space station, human genome sequencing, and similar projects would freeze out more meritorious, smaller, more promising science. The difficulty, they continued, lay with scientists themselves. They refused to take a collective stand or to make distinctions between the merits of several programs. Instead, individual scientists offered their opinions, a fact that was causing science to lose its reputation as "balanced, fair, and analytical." "Our internal dissension and the mixed, conflicting, and self-serving advice," Press argued, "are threatening our ability to inform wise policy making." Rosenweig maintained that no scientific group is "prepared to participate constructively" in science budget matters. A "total absence of help for any seriously minded government official" wanting to think about science has produced a series of "uninformed and dangerous decisions by government."

Press wanted "expert advisory groups" to set policy. He despaired that "the President's policies set the country's scientific priorities." Press personally thought the "highest priority" in the next administration should be given to providing research grants "that reach the largest number of scientists, engineers, clinical researchers, and graduate students" and to resolving "'national crises' as the AIDS epidemic and the loss of the nation's space launching capacity, and research to exploit 'extraordinary scientific breakthroughs.'" Clearly, Press's sensibilities differed from the Reagan administration's but he offered little more than that. He opted for the technocratic vision of a self-defined scientific elite guiding America as the alternative to the messy business of electoral politics. He was soon joined by Hornig, Johnson's science advisor, and William Golden, who put together the science advisory group for President Truman.[34]

But not all former presidential science advisors saw science policy recommendations similarly. David, Nixon's science advisor, complained that scientists could not keep their ideological convictions separate from their technical advice and so offered opinion rather than substance. Their advice needed to be viewed with that caveat in mind. Stever expressed suspicion over any centralized arrangement, such as a strong science and technology office, because it spoke with one voice. He preferred a decentralized system where presidents received a plurality of views and weighed among them.

Science societies, especially the AAAS, took up the matter. Members resolved to publish white papers to guide the candidates. They also recommended that George Bush and Michael Dukakis pick and announce their science advisors before the election. In that way, they could have the benefit of their advice when filling all other governmental posts, many of which touched in some fashion on science or technology. To demonstrate how important they thought the science advice question to be, they invited the two presidential candidates to a forum to discuss the matter. Both *Science* and the *Washington Post* published the two candidates' views.

Claiming "that there is virtually no aspect of government that does not involve science and technology," Bush favored continuation of all the Big Science programs begun under Reagan, a doubling of the NSF budget over five years, and additional money for Little Science. He cited a Republican Party plank that pledged "to reinforce" the White House Office of Science and Technology. Candidate Dukakis also supported Big Science, doubling the NSF's budget and expansion of Little Science funding, but he wanted to slow the rate of increase in SDI research. His proposal for science and technology in the Executive Office was more detailed than Bush's and he promised to involve it in all budget decisions. In either case, it appeared that the Reagan program would continue in very much its present form into the next administration.[35]

Chapter 4

OSCILLATIONS AND PERTURBATIONS THROUGH MANIPULATIONS

Rise of the New Politics

From Reagan to Bush

I n the last years of the Reagan administration, the Defense Department accelerated its efforts to develop federal-industrial research connections in a series of cutting-edge technologies. Through the Defense Advanced Research Projects Agency (DARPA), the department focused on potential dual-use technologies—those that would have both military and commercial applications. Semiconductor manufacture, high-temperature superconductors, neural networks, and high-definition televisions all flourished as DARPA projects. The Defense Department's technological vitality did not escape notice. No less a notary than Robert White, president of the National Academy of Engineering—the NAS's engineering complement—congratulated the department in the early months of the Bush administration for becoming "the nation's de facto Ministry of

Technology and Industry." Competitiveness had yielded an "emerging national consensus" on the need for closer links between government and private enterprise "to develop commercially important technologies" to combat Japan and Europe. We "spend our time debating only science priorities," White contended, when our technological ones are no less critical "to secure our industrial future."[1]

White demanded that the Bush administration extend the industry-government partnership. But the kind of mini-MITI established in DARPA was not the only approach that the pioneering Defense Department employed to counter Japan. Just as it had since it first employed banks of psychologists, sociologists, and economists to spin out scenarios to anticipate military matters in the 1960s, the department turned to social science. It institutionalized in all Defense Department activities a version of the Japan-inspired management techniques that collectively had become known as Total Quality Management.

Those very same management techniques had been memorialized in the creation of the Baldrige National Quality Award a bit earlier. To many of TQM's proponents and devotees, its ability to raise productivity markedly made it the key to winning the competitiveness battle. And the Defense Department valued winning more than any other federal agency. Winning was its mission.

No part of the federal bureaucracy seized this new managerial ethos more passionately than the Defense Department. In late 1988, at the department's behest, the Reagan administration designated TQM as the "major strategy" for all defense activities. By 1989 the army leadership proclaimed it had completed "the transformation to a customer driven culture" in all army activities.

The army's Communities of Excellence program, established as "customer driven" during Bush's presidency, brought quality management to every army endeavor. None was more stellar than the Redstone Arsenal, which specialized in missiles, rockets, and other long-range weaponry. It received commendations from the program's inception. According to the arsenal's mission statement, "at the heart" of the arsenal's work on delivery systems for weapons of mass destruction "are individuals committed to the success of this program." The communal ambition was to "achieve performance excellence everywhere in people, services, and facilities" through "a shared vision" and "clearly stated goals." Only those techniques provided "employees with an understanding of their organization's constancy of purpose," which allowed Redstone "to provide continually improved products and services to its customers and a rewarding environment for its employees."

The Redstone vision of what its missile systems are expected to do clashes with these communal, noncompetitive sentiments. The vision's three components are "decisive victory through excellence in missilery," "power projection through superior technology," and "consistently exceeding customers' expectations" (the army is the sole "customer"), each of which depend on "an inspired workforce committed to excellence." Its goals include to "achieve world-class customer service," "generate the science and technology to enable America's army to swiftly achieve decisive victory," and "empower and train the total workforce and provide them with a quality work infrastructure and environment."

Goals and shared vision must be reached in accordance with Redstone's core "values": "integrity," "competence" defined as "giving your best, nothing less," and "leadership" through "enabling others to use their talents and resources to excel in all tasks." Other core values are: "loyalty," which is "giving constant support to the organization, its people and its customers," while at the same time giving "respect for the needs and contributions of all"; "commitment," identified as "willingly, consistently carrying out goals of the organization" that is, "excellence driven [and] customer focused"; "teamwork" as "empowering people, sharing leadership and cooperating with one another" as well as "respect for all team members"; and "service" that exceeds "our customers' expectation in a professional, pleasant and competent manner."[2]

TQM soon spread through the federal bureaucracy. A customer-driven ethos, vapid mission statements, and an explication of core, humane values characterized the new tool to manage away any semblance of conflict or dissent. NASA's John F. Kennedy Space Center developed its first strategic plan and quality control procedures under

Reagan and incorporated TQM in toto under Bush. The Central Intelligence Agency's Office of Information Technology adopted TQM in 1991. The National Science Foundation, the National Endowment for the Humanities, and the Department of Education all were long-time devotees to Total Quality Management practices. Each began their implementation during the early years of the Bush presidency.

Emphasis on management, on placing the appropriate factors and activities together, and on organizing them in a most efficient fashion were marks of the Bush administration. Despite occasional rhetoric otherwise, Bush lacked the ideological commitment to zero-based government that Reagan had expressed. His style and emphasis were more managerial, somewhat reminiscent of the Eisenhower administration, where managing challenges rather than reorienting the federal bureaucracy mattered. Perhaps Bush adopted a managerial approach because he had come to the White House with significant managerial experience. He had served as ambassador to the United Nations, de facto ambassador to China, CIA director, and Republican National Committee chairman. Rather than innovate or launch bold new initiatives like his predecessor, Bush sought to bring the appropriate constellation of interests and forces together and to have them arrange a resolution to a particular problem or establish a policy. John Sununu, an engineer by training and the former governor of New Hampshire, served as Bush's chief of staff and rode shotgun on the management process to ensure it occurred as effectively as possible.

Uneasy Rests the Mantle

It was that attention to structure and propriety that led Bush to promise to create in his administration a vibrant, active Office of Science and Technology Policy. Bush certainly had help in deciding to raise the profile and responsibilities of the science policy office. In an unprecedented move, the National Academy of Sciences, the Institute of Medicine, and the National Academy of Engineering joined together to call for Bush to restore the policy office to its former luster. They objected to the Reagan administration's refusal to place its faith in the hands of the science establishment, choosing instead to chart its own agenda. This particularly offended NAS president Press. Identifying it "as a sign of the times," Press thought it essential that the academies move proactively and provide advice to the Oval Office on a regular basis without executive request. In effect, the academies would single-handedly and without partisanship set the agenda for national science and the revamped and empowered White House science office would shepherd the policies into fruition. To show that they were ready for the challenge, the academies identified four areas that demanded critical decisions within the next four years. AIDS, the global environment, and space were the first three, but the fourth was hardly a surprise: creation of a large, high-profile science policy office to channel direction from the academies throughout the federal bureaucracy.[3]

Placing a science office that high in the academies' priorities demonstrated just how embittered they had been with Reagan's approach. It was not about money and less about power than respect and with it some autonomy. The other areas the academies highlighted were mundane, unexceptional; it was not about holding government's feet to the fire to have it move in new directions. They felt disrespected and sought a mechanism to ensure that they received the influence they thought themselves entitled to. In a sense, they claimed Reagan victimized them and hoped to empower themselves through bureaucratic arrangement.

The managerial Bush shared their science office concerns. He had initially planned to select a science policy director before he took office so that the new person could advise him on subcabinet-level appointments in highly technical areas, such as energy, agriculture, and commerce, as well as coordinate the federal science effort. But he refused to rush to select his new science policy director. It was almost May when Bush finally announced his choice. The delay deeply disappointed and angered establishment science. To them, it seemed as if Bush was continuing the Executive Office's callous disregard for formalized scientific advice, which they recognized as a rejection of science generally.

Bush's appointment, Allan Bromley, had impeccable scientific credentials. A Yale physicist, he had served on NSF, DOE, and NRC advisory boards. Bromley also held impeccable Republican credentials. He had close research ties with IBM and AT&T, giants in American corporate research and development. Perhaps as important, Bromley was a personal friend of Bush. To induce him to join the administration, Bush placed the science advisor post on par with the national security advisor, a single step beneath cabinet rank, and granted him direct access to the Office of Management and Budget on all budgetary issues. The administration also allowed Bromley to plan for office expansion. Staffed by less than a dozen scientists under Reagan, Bromley expected to employ up to fifty staffers in policy office posts.

Long on record as favoring a discrete, consistent, formal American science policy, Bromley immediately faced the challenges of rationalizing science policy, managing competing entities, and structuring the federal science and technology effort. And he did so in a characteristic Bush administration sort of way. Bromley brought together the interested stakeholders and sought to negotiate among them while at the same time insisting on the parameters to which any resolution must conform. But this complicated procedure and complex agenda were made even more difficult by a controversy that had begun just before he was appointed. A Utah chemist and his English collaborator announced in a press conference a possible discovery that rocked American science. They claimed to have produced a fusion reaction at relatively low temperatures through chemical means. If their contention were true, it meant that America's energy problems would be solved for the foreseeable future.

But it also exposed various tears in the fabric of American science. By choosing to announce their discovery in a non-peer-reviewed medium, they circumvented any scrutiny by their peers, a long-standing tenet of modern science. They further sought to skirt peer-review by demanding immediate funding from Utah as well as Congress for their work. That these investigators were chemists also was an issue. Fusion studies had long been the province of physicists, who pursued grand strategies with incredibly expensive equipment. Here the chemists suggested that money and effort had been wasted; fusion was attainable not by the staples of the science of physics—high temperatures and magnetic bottles—but with common chemistry laboratory apparatus. Finally, the chemists' announcement seemed a critique of big, expensive, theory- and machine-driven science. The idea that large-scale projects were the most fruitful way to develop new insight blinded scientists and their benefactors to the recognition that bias toward large-scale projects was not only expensive but also constraining; it eliminated projects that might prove more profitable simply because they failed to incorporate giant machines or complex environments. It inhibited scientific vision. The seamier side of science also appeared in the cold fusion imbroglio. A professor at MIT filed a series of patents around the project—apparently MIT encouraged him to seek priority—even as he acknowledged that he did not believe that the process actually produced a fusion reaction. Corporations from five continents contacted the Utah scientific group and tried to cut deals with these men to control whatever discoveries might emerge.

Announcement of cold fusion created such a stir that thousands of investigators immediately turned their attention and federal funds to confirming or disproving the experiments. Initially, results were mixed. Some experiments seemed to support the chemists while others seemed to refute their work. Bromley stayed on the sidelines but the Department of Energy sponsored a three-day conference to discuss the conflict. Determinations announced there were overwhelmingly negative. A panel of governmental scientists looking into the matter a month later urged the government to refuse to fund cold fusion experiments. Arguing that the experiments so far were "not persuasive," its members proclaimed themselves "pretty unanimous" in their recommendation. Utah continued to back cold fusion work but the federal government discontinued support. Yet three months later, members of an NSF workshop argued that most recent work suggested the additional federal funding would be desirable. That recognition caused the DOE to reconvene its panel to reconsider its earlier recommendation. While acknowledging that cold fusion raised "some questions of scientific interest," the "present evidence for a new nuclear fusion process is just not persuasive."[4]

As the debate over cold fusion raged within American science, Congress, which the Reagan administration had restored to relevancy, held hearings on scientific misconduct. In particular, it demanded a specific program to treat misconduct, conflict of interest, and other ethical violations involving federally funded research. What was up for debate was who should investigate and regulate misconduct. Should it be the agency that funded the research, the institution that employed the researcher, Congress through statute, or the scientific community generally?

This was the context in which Bromley accepted his post. As part of the elevated stature of the position, Bromley needed to undergo senatorial scrutiny and confirmation. The four-hour hearing was far from friendly. Senators grilled Bromley on the environment, on Big and Little Science, and on what science budget cuts to make in wake of the deficit. Almost to a man, senators urged him to adopt a more forceful role in federal science policy. On that issue, Bromley agreed. But much to the dismay of some senators, especially Al Gore (D-Tennessee), who chaired the meeting, he pronounced himself not yet convinced of the reality of global warming or that the nation needed to cut its carbon emissions. Bromley wanted to continue the entire Reagan Big Science agenda but also argued that Little Science needed enhanced support. Little Science, he maintained, was "the backbone of our science research base and major science initiatives." Gore suggested that the nation lacked funds to support the entire Reagan science program and asserted that defense research and development should be cut. Bromley's rejoinder was to argue that defense R&D would likely have to increase to counter the growing Soviet threat. If anything could be reduced, Bromley favored cutting annual support for the Human Genome Project and extending its years to completion.[5]

Bromley pushed each of those points during his first year in office. He also lent his support to a project championed by Gore a year earlier. This five-year plan called for creation of a large federal interagency kitty to research and apply high-speed computing technologies. The result would be a doubling of federal money spent in this field. Part of this new fund would go to create a network sixty times faster than any previous data transmission network, enabling researchers studying such diverse projects as the human genome, superconductivity, and semiconductors, to share huge quantities of information.

Bromley himself pushed a similar five-year plan for superconductivity. Indeed, the idea of a five-year plan itself was a management tool. It set a precise and concrete agenda and direction for what in science was an extremely long period and thus guaranteed where efforts would be undertaken and where results would reside. This sort of planning fixed both the present and the future because it privileged some possibilities and removed others. Privileging certain approaches and goals reduced waste in that it refused to support activities outside those areas but it also circumscribed where things would go. For those who agreed with those outcomes, planning was a success. For those whose interests lay elsewhere, their future had been managed out of existence.

In the case of superconductivity, Bromley would increase federal expenditure by more than 50 percent in five years, but that and much already-budgeted money would go not to new research but to translating federally sponsored research into commercial applications and industrial products. In this and other cases, the Bush administration marked a sharp break with the Reaganites in that they saw federal dollars as crucial to the implementation of science; government should occasionally fund development of commercially viable products as well as the science that undergirds them.

That was a huge concession and it cut at the heart of the Reagan agenda. It suggested that totally free markets were not the best way to manage things but that the federal government could periodically direct where American industry and business ought to go. Yet there were boundaries that the Bush administration refused to cross. It repeatedly objected and fought against all attempts to create publicly supported corporations to undertake some task that it felt remained the province of private enterprise. Government could initiate a process or uncover a technique but it must not create a sustained publicly funded corporate structure to regularly seek to dominate a field.

Bush himself fashioned part of the science management framework when in early 1990 he appointed a twelve-person science advisory committee. Headed by Bromley, it was composed primarily of two types of persons. Some were experienced in translating science into applications, such as Norman Borlaug, who won the Nobel Prize for his part in the Green Revolution using scientific concepts, and chemical and biological inputs to increase agricultural yields in third- and fourth-world countries. Others had a direct stake in the commercial products of science, such as David Packard, chairman of the board of Hewlett-Packard, and Solomon J. Buchsbaum, senior vice president of technological systems at AT&T Bell Laboratories. This committee, which reported directly to the president, embodied the sort of management strategy favored by Bush. He assembled the players around an issue and worked with them to devise an approach that each could embrace.

Bush found several other ways to show how intensely he valued science. He placed Bromley and the entire science and technology policy team in the West Wing of the White House, symbolic of the epitome of political access. He also became the first president since JFK to address an annual meeting of the National Academy of Sciences. There he repeatedly genuflected before science. Claiming he came not "as an expert but as a believer," he spoke of "the magnificent creativity and energy of the American technical community" and thanked scientists for bringing the "beauty and grandeur of this world . . . into sharper focus." Bush pledged to accelerate support for science and took care to thank the assembled for their making "science and technology . . . forces for change." Here he meant that the new elements of communications—faxes, satellites, VCRs—had become "a potent force for peace"; "pictures from Poland" and "scenes of the Berlin Wall" had been "more powerful than chisels for breaking down barriers" and "etching the idea of freedom on the psyche of humanity." The future seemed even brighter. "Now more than ever, . . . government relies on the impartial perspectives of science for guidance." Bush referred to his revamped science policy enterprise as well as efforts to ensure that federally supported science diffused throughout government and private enterprise. He promised to redouble government efforts to fund scientific initiatives.[6]

Bush proved a man of his word when he sent Congress his science budget. It retained all the programs created and supported by Reagan—Bush's budget accelerated funding for these Big Science initiatives—but it also included hefty increases in the basic research budgets of Defense, the NIH, and the NSF. New agricultural science research programs also were on tap as were NASA satellites solely to track potential global environmental change.

Some scientists hailed Bush's science budget as a breakthrough, while others thought it acceptable in troubled economic times. But a considerable segment of the scientific community recognized the budget as a disaster, as hampering America's scientific and economic future. They argued that every year fewer and fewer scientists received grants. Although funding was dramatically higher, the size of grants on average far outstripped the pace of funding growth. Moneys for the Super Collider, superconductivity, NASA, the Humane Genome Project, and other Big Science items reduced funding for normal and Little Science. Members of the American Society for Cell Biology and the American Society for Biochemistry and Molecular Biology were so distressed by their prospects that they went so far as to hire their own lobbyist to secure more money for their constituencies. "Mom-and-pop operations are the future of science in our country," their representatives proclaimed. Already changes were occurring. Less than 30 percent of the applications submitted to the NSF and the NIH received funding. Young people will "see the handwriting on the wall" and not chose science as a career, they contended, because there will be no grants for them when they graduate. As a result, the nation will in a few short years be unable to compete in the scientific marketplace. George Chapline, senior physicist at a federal lab, flatly stated that "almost without exception," advances "resulted from work done by a few individuals working in relative obscurity—not as the result of a huge government project. The wellspring of science as well as the practical spin-offs from science is the solitary researcher."

To some investigators, Big Science caused the epidemic of pork barrel science that plagued federal initiatives. As huge projects drew funds that might have gone to support peer-reviewed grants, the number of competitive awards dropped precipitously. Demographic data indicated that more than "87 percent of all scientists and engineers who had ever lived [were] active today," and universities and their faculties argued that they had no alternative but to circumvent the granting process, hire lobbyists, and go directly for congressional earmarks. As science was being subjected to "ration-coupon conditions," "real world pressures" led them to "all varieties of manipulation of science's business."[7]

Others argued that Big Science would inevitably lead to ethical violations. As small scientists found themselves increasingly strapped for funds, the temptation to pronounce their findings prematurely or through the media rather than peer-reviewed publications would be overwhelming. Some even suggested that desperate researchers might fabricate findings simply to ensure continued federal support.

Still others saw social policies as hindering pure, apolitical science and freezing out bona fide scientists. They claimed that recently the bulk of NSF money had "gone to bolster science education, to give minorities and women greater access to scientific careers" rather than to support good science. Still others contended that grants had not simply become bigger but also were offered for more years because of "sophistication inflation," where each new discovery in a field cost orders of magnitude more than the previous breakthrough. But in the most sustained comments surrounding Big Science projects, Rustum Roy, Penn State materials science professor, made the case simply. "Big science has gone berserk," he said. Many "good minds and a lot of money are going into areas that are not relevant to American competitiveness, American technological health, or even the balanced development of American science." The result was a catastrophic "Two Cultures" problem. Big Scientists don't understand why Little Science enthusiasts "do not share their excitement at the prospect of grand discovery—triumphs of the human spirit—no matter the cost." Little Science "researchers are offended" that scientific colleagues would selfishly seek projects "so huge" that they "must inevitably suck resources away from other fields."[8]

The Bush administration sought international partners as a way of defraying some of Big Science's costs. Japan was a natural target, long having benefited by constructing commercial applications from American science. The administration sought to have the Japanese channel some of the money they would have used to make commercial products to help fund the American science upon which those products were based. The Superconducting Super Collider was a focus, but James Watson carried the matter a step further. In a series of menacing and otherwise undiplomatic messages, he demanded that the Japanese "pay their fair share" of Human Genome Project costs or be banned from employing any of the research. The resulting international incident did little to resolve the fundamental issue.

Watson's ham-handed approach was unusual. While virtually all partisans in the United States understood that political action was the inevitable result of tensions within science, they also understood that support for science rested to a large degree on the appearance of being above the political fray. Each time they took their case to the people or their representatives in Congress, they tended to demystify science, to reduce it in the public estimation to just another animal feeding at the federal trough. In short, they argued for scientific exceptionalism. They sought additional funding through any means by arguing that science and technology had become essential components in the public weal; they were not a special interest but rather the public interest. No matter what their particular cause, they tried to paint a nuclear winter scenario in which failure to support whatever they wished would inalterably ruin American life.

Leon Lederman, a Nobel Prize–winning physicist, embodied the doomsday thrust. As incoming AAAS president, he conducted a survey of scientists to determine the "quality of life" of the enterprise and sent out his report and conclusions to the organization's 130,000 members. Titled *Science: The End of the Frontier* in juxtaposition to Vannevar Bush's *Science: The Endless Frontier*, Lederman found "deep concern, discouragement, frustration, and even despair and resignation." Scientists complained that seeking funding took an increasing amount of time, that regulatory requirements and increasing overhead rates gobbled up their grants, and that they were "embarrassed because they feel they are unable to serve as adequate role models for their graduate students." Lederman cited the myriad contributions of science to the American economy but also noted "the profound cultural . . . benefits science brings." Without international scientific leadership, Americans would lack "self-confidence." Scientific leadership "inspires our youth, creates a sense of endless frontiers of the human mind and of human aspirations." Loss of "scientific and technological exuberance" would be "the greatest penalty in the long run for the decline of the research system."

With America's future at stake, Lederman rejected the notion that by raising these points he was appearing self-serving, bringing science down to the level of "just another interest group." Since "US science was at risk," he felt there was no other course to take. He cited defense spending as what could happen to science funding in times of national need. During Reagan, America doubled its defense spending, raising the total to more than $300 billion yearly, even as the nation experienced economic distress. A similar total in science would restore the endless frontier. He called for a special commission composed of White House, congressional, financial, industrial, and university representatives to devise means to generate increased revenues to be used exclusively to expand federal science support. One of the methods that Lederman advocated was a special national tax to establish a research trust fund. He also mentioned that revenues from federal saving bonds should be diverted to fund science. Anything less was to court disaster.[9]

Not all of Lederman's science colleagues agreed with his logic. "Deep down all my colleagues believe the feds owe them a living," noted Roy. "They are nothing more than 'welfare queens in white lab coats.'" "Just because you get a PhD," he continued, "doesn't mean you're entitled to federal support." Aware that he might be committing "treason in academic circles," Erich Bloch, former NSF director, agreed. "Scientists," Bloch claimed, "have no unalienable right to funding."[10]

CONGRESS VS. THE PRESIDENT

Responding to doomsday scenarios never was easy. Politically that task rested in Congress. With its budgetary authority, Congress made choices. Certainly Big Science had big constituencies. Little Science had no easily identified corpus of supporters. Both groups as well as others expected a "peace dividend"—less military spending—now that the Soviet Union had collapsed, but the Bush budget citing deferred projects actually called for a 3 percent defense increase. The cudgel of the Gramm-Rudman-Hollings Act hung over the entire process. If the deficit reached more than $65 billion, then Gramm-Rudman-Hollings mandated across-the-board cuts in discretionary funds to reduce the deficit beneath that figure. Since discretionary funding comprised a small part of the federal budget, invoking Gramm-Rudman meant significant cuts in all science programs as well as other facets of the social safety net. It placed Congress in an awkward position. By authorizing everything, they would balance the budget on the backs of science and social welfare. To try to keep the deficit under control required them to prune budgets selectively, to make hard choices rather than automatic across-the-board cuts. They could choose the enemies that they would make rather than anger all constituencies.

Congress generally adopted the latter approach but they also used accounting and parliamentary tricks to keep Gramm-Rudman-Hollings at bay. Presuming from the position that the federal government was not "in business to make sure that every PhD gets funds," Congress picked its winners and losers, although in most cases, the losers merely got a reduced appropriation. Yet that failed to quiet supporters of certain causes. For example, discussions to reduce appropriations for the Superconducting Super Collider for one year led Lederman to suggest that if he were in that field, "I think I'd slit my wrists." Budget reductions, while so close to discovering the "top quark," a building block of a form of matter, was "snatching defeat from the jaws of victory."[11]

The Super Collider was not alone. Congress funded human genome research at about 65 percent of what Bush requested.

Increasingly, however, Congress accepted an offer by the NAS's president, Press. He urged Congress to have scientists rank the importance of scientific projects and he volunteered the NAS's assistance. Congress commissioned fully twenty-four NAS studies during 1990, up nearly threefold from recent years. Congressional staffers began to take monthly lunches with academy members and members increasingly had weekend retreats with congressmen.

The NAS's increased role sharply conflicted with the Congressional Office of Technology Assessment. In March 1991 it unveiled its new study of the health of the American scientific enterprise. In direct repudiation of Lederman and others, it found the present period to be science's golden age. Several factors seemed critical. First, research expenditures nearly doubled in the previous two decades but the number of scientific papers more than doubled; scientists now on average were more productive. Second, despite the tumultuous cries to replace the scientific infrastructure, science administrators at the top fifty universities rated their research facilities "good" to "excellent." Third, while it was true that a lower percentage of grants were funded than before, it was also true that more scientists now were supported by federal funds than ever before; there were simply many more scientists now. Fourth, five states received more than 53 percent of all federal science financing, yet another explanation why universities and states scrambled to collect earmarked funds. Finally, it disputed the contention that scientific bodies should select areas of inquiry to fund. Peer review only worked within disciplines; it failed across disciplines because there were no interdisciplinary peers. The result would simply be politics, tradeoffs, and negotiations among scientists. Congress could sort through various proposals much better, it concluded, because scientists lacked the perspective and ability "to judge issues such as timeliness and the social and economic benefits of competing projects"; they were scientists, not popularly elected representatives.[12]

As Congress mulled over the conclusions of its Technology Assessment Office and the advice of scientists generally, new immediate scientific concerns emerged. Distinctive research into medical conditions peculiar to women became a potent topic in the late

1980s. As early as 1986, the US Public Health Service noted that in some instances women's anatomy and physiology differed so markedly from men's that both sexes needed to be considered in most medical testing. The NIH quickly incorporated this sentiment in its suggestion that researchers included both sexes in their studies. In wasn't until 1990, though, when that stipulation made it into legislation. The General Accounting Office reported that few scientists had chosen to follow the NIH guideline and the Congressional Caucus on Women's Issues sponsored a bill to force federal grantees to include women in their experiments. Quickly the NIH wrote this requirement into its grants and even established a new Office of Research on Women's Health to chart medical issues unique to women. Bernadine Healy's appointment as NIH director accelerated women's health research. As the first female head of the major federal medical research agency, Healy quickly announced a $600 million program to redress gender imbalance in medical research, thereby permanently institutionalizing women's medical issues as separate and different from men's.

Healy's program delighted women's health activists, who redoubled their commitment to raise women's health issues. For a series of cultural and medical reasons, breast cancer proved a focus. Agitation and publicity led a politically savvy Congress to fund research and treatment of the disease much more generously than it had in the past. In 1990, Congress appropriated $90 million; in 1991, $135 million; and in 1992, more than $400 million.

The AIDS plague of the late 1980s received similar attention. Here the issue was regulatory. As protector of the public health, government through the FDA had long demanded scientific certainty before clearing a drug for the market. That meant a series of double-blind tests on populations of significant size over the course of years; as each initial step was satisfied, the size and diversity of the population being tested expanded. As a consequence, it took as long as ten years for some drugs to secure approval and reach the market.

With AIDS, ten years was more than an eternity. There was no cure in the late 1980s and little in the way of relief. AIDS activists, many of them gay, organized to dramatize the plight of those with the disease. Partnering with members of the media to draw attention to the cause, activists focused directly on FDA regulatory policies. At the heart of the matter rested several conflicting obligations: the FDA's responsibility to protect the public's health from adulterated or dangerous products; the scientists' responsibility to test materials thoroughly and in a scientific fashion so as not to risk the well-being of populations and to conclusively prove a certain result; and the lives of persons with AIDS. That AIDS seemed unfailingly fatal helped tip the balance. Conventional remedies meant certain death. Suspect remedies or remedies that had not yet cleared all safety hurdles could do no worse. Recently appointed FDA head, David Kessler, announced the new compromise. For diseases that were inevitably fatal, Kessler proposed in 1991 a policy of "accelerated approval." In practice, AIDS drugs under Kessler were approved in about half the time of other medicines.

Some policy questions resisted congressional and scientific adjudication. Global warming remained contested throughout the Bush years. Environmental organizations, presidential advisors, and congressmen, especially Gore, repeatedly called on the Bush administration to reduce greenhouse gas emissions. At first, the administration refused, citing an indeterminate science. That had been Bromley's position. But even after evidence seemed to indicate that global warming was likely a real phenomenon, and even after the NAS advocated aggressive energy conservation methods such as a "planetary insurance policy," the administration still balked. Here Sununu raised economic questions and competitiveness became the critical issue.

As under Reagan, the Bush administration feared that American industries were at a disadvantage in the international marketplace. Again two major culprits emerged. Critics pointed to American regulatory/tax policies as one factor. Japan's MITI remained the other principal villain. In the case of regulatory delay, Bush adopted Reagan's program of establishing a competitiveness council headed by the vice president to oversee regulatory/economic conflicts. Under Vice President Dan Quayle, the council moved behind the scenes, singling out regulatory action in an agency and pressuring the federal operative to modify the rule in pursuit of international competitive advantage. The aggressiveness with which Quayle's group moved toward that mandate angered Congress and many watchdog groups. Repeatedly it was accused of substituting political philosophy for enacted legislation. As critical, some of its members seemed to rule on regulatory laws in economic sectors where they held a fiduciary interest. The conflict of interest claim became widespread in late 1991 and tempered council activities.

Direct anti-Japan/MITI maneuvers were varied. On one hand, the Bush administration expected Japan as a primary beneficiary of American scientific activity to pay a share of research costs. Watson's Human Genome Project screed was one example but the administration itself made its demands known on the highest levels. Funding for the monstrously expensive Super Collider was an administration priority and Bush traveled to the island nation to make the case directly to the prime minister. Unfortunately, the flu-ridden president failed to keep down his meal as he turned to talk to the prime minister. Shortly thereafter, Japan refused to support Super Collider research.

Attempting to mimic MITI was more common, although Republican free-market principles made any such approach questionable. Bush refused to pick economic areas in which to invest. His corporate consensual approach would not let him select areas to favor or that required federal assistance. In fact, a high-ranking technology official quit the administration in early 1990 over Bush's refusal to establish a full-blown high-tech support plan. But he did develop a strategy of using federal funds to support only what Bromley termed were "generic, pre-competitive technologies." Arguing that economic "competitiveness and national security aren't separate anymore," the administration listed the new high-speed data transmission super-highway as a prime example of that emergent technological type. In March 1991 the White House expanded its "generic, pre-competitive" list to include such things as inexpensive, flat screen monitors, x-ray-based lithography, and procedures to increase precision machine tool accuracy. The administration also inaugurated plans to shape the national laboratories "to better collaborate with American business." We need to get, Bromley maintained, "people from industry more involved in decisions that are now entirely internal." We can no longer afford to permit lab scientists to do "things just to satisfy their own curiosity."[13]

In April the White House published an expanded list of key technologies that required "increased cooperation between government and corporations." Each was "essential to American economic and military strength." Reference to military strength was intentional. The nation's citizens had just witnessed more than a month of precision bombing in Iraq and a swift battle—Operation Desert Storm—where America's superior technology had enabled its forces to destroy the world's fourth most powerful army with fewer than one hundred American fatalities, many of them from friendly fire. The revised group of technologies announced by the administration included software, pollution treatment equipment, aeronautics, and new lightweight materials for planes and military vehicles. These areas were designated "critical technologies." As a further demonstration of the new thrust, Bush created a Critical Technologies Institute and placed it directly in the Office of Science and Technology Policy.[14]

Almost immediately Democrats contended that Bush's support for some technological sectors constituted an embryonic national industrial policy and in the papers and through Congress they urged expansion of the program. The American Technology Preeminence Act of 1991 came from that sentiment. It went so far as to provide federal loans to embattled technologies so that they would not collapse in America or relocate overseas. "Too long," noted Congressman Richard Gephardt (D-Missouri), "we have failed to compete . . . in the area of commercializing technologies." But while Gephardt maintained that business alone lacked the mettle to fend off foreign competitors, administration members objected to the bill "improperly" funding "particular product-development projects" instead of "broad technology goals." Bush himself threatened a veto.[15]

A national technology policy would emerge as a critical factor in the 1992 election. Bush's failure to fulfill his campaign promise to provide thirty million new jobs during his presidency—only a million were created during his first three years—proved a potent reminder of the draining of technological possibilities to foreign places. But well before the election took place, Congress asserted itself in a new way. It insinuated itself right in the middle of science funding and policy decisions.

From the inauguration of the NSF in 1950, Congress participated in funding questions in a rather circumspect way. It nibbled here, refused to fund a project there, objected to certain social science research. But it accepted as right peer review and even as they circumvented it with direct appropriations they generally did not tamper with the awards given out in the more conventional manner. That de facto consideration underwent a fundamental reevaluation in the last years of the Bush presidency. Congress incorporated the essence of the rhetoric that science and technology was the basis of America's future and so questioned every premise that came before it. As significant, Congress recognized its relationship to the White House in a new way, not as the loyal opposition but rather as an alternative strategy. Rather than agree on principles and disagree on methods, Congress and the White House both decided that it was within their rights to formulate and implement policy. This generally broke down according to partisan political lines, of course. Those whose party held the White House favored presidential authority, while those in control of Congress backed congressional power.

It was the unwillingness to cede one to the other that marked the early 1990s. Both bodies and parties acted as if they alone had the successful strategy. Either you held to one philosophy or the other; there was explicitly little willingness to seek middle ground. In early 1991 Congress took an especially active role in reshaping Bush's science budget. It dramatically changed the appropriations for such projects as the Superconducting Super Collider and NASA's space station, but it also debated the possibility of standardizing lowered overhead rates on all federal grants. Congressmen saw as outrageous overhead rates proposed by universities in excess of 100 percent—half the funding went to the grantee, the other half to the institution—and proposed to reduce rates significantly. Universities objected, especially at the prospect of congressional activism and arbitrariness. Carol Scheman, speaking for the Association of American Universities, which represented most of the research universities in Washington, put the matter bluntly when she said, "There is no such thing as a legislative solution that's a good idea."[16]

Congress's plan to reduce the scientific and support staff of Department of Defense labs by as many as fifteen thousand research jobs over five years seemed an even more draconian restructuring of the pursuit of American science. By threatening to eliminate those posts, Congress was in effect proposing to reduce the single most prominent technology-industry connection, DARPA. The move seemed antithetical to the thrust of both the Democratic and Republican parties, each of which hoped to increase American competitiveness in global markets.

But the most profound scientific tumult was a direct manifestation of the 1992 election. With domestic spending the issue, Bush attacked Congress's involvement in setting scientific priorities and constituent services by identifying sixty-seven earmarked projects totaling about $30 million as pork barrel science and by demanding that Congress excise them from the budget. Pledging that this was only the "first in a series" of cutbacks, Bush promised to embarrass each of the measures' congressional sponsors if the body did not heed his ultimatum. The administration put the case clearly. These projects were "not selected on merit or peer-review basis, are not nationally significant, and would not have qualified for federal funding in a traditional competitive process."[17]

Congress responded in kind. Led by Senator Robert C. Byrd (D-West Virginia), it flagged thirty-four peer-reviewed proposals and refused to fund them as retaliation for the Bush administration's bold gesture. His justification was that these funding requests did not encourage American competitiveness, lead to other economic development, rest in the area of critical technologies, nor produce a new basic understanding of fundamental science. Scientific organizations, including the American Psychological Association and the Council of Scientific Society Presidents, vehemently complained of Congress's action. "A political battle is being fought on the backs of scientists," one chemist noted. His discipline's chief lobbyist sounded an even more desperate note. "This time it's the behavioral scientists," remarked Kathleen A. Ream, but "next time it could be our scientists being used as a political football." As a consequence, "our near and dear" peer-review system "could go right down the drain," argued University of Washington vice president Jack Lein. Science sounded the call to arms. "An attack on the social sciences is an attack on all science and should be rejected by the scientific community."[18]

Both houses of Congress then passed without emendation each of the measures that the Bush administration demanded that they cut. But while the legislation did not explicitly require Bush to reject the proposals that Byrd had highlighted—the bill did provide what it called guidance to the NSF and mentioned the grants by name— it did reduce the NSF budget by exactly the amount of those grants, about $2 billion.

The act went further. It compelled the NSF to emphasize research "focused on the fundamental laws and systems of science that supports the nation's technological base, that supports the nation's economic competitiveness, and that improves the nation's mathematics and scientific base." As important, the measure required the NSF to review its procedures and explicitly report to Congress how it was going to ensure that no taxpayer money went to fund projects that did not fit under this new congressional umbrella.[19]

At roughly the same time, Congress undertook another series of investigations that likely would divide the scientific community. It looked into the condition of women in male-dominated fields, including research science. Overt sexism and harassment were reported, although witnesses maintained that it was less pervasive than in the past. What seemed more immediately relevant was that the scientific workplace discriminated against women. Women had maternity and childcare concerns, and most universities and corporations failed to account for them. Women received lower pay than their male scientific counterparts and a glass ceiling seemed to restrict their rise.

Attention was diverted from the women in science question when in early July the House refused any funding for the Super Collider. Although the Senate later voted to restore support, the future looked questionable. Texas—where the Super Collider was to be built—and Louisiana senators lobbied their colleagues hard and brought in numerous scientific luminaries to testify to the project's supreme intrinsic scientific merit. NASA, the NSF, and the NIH proved less fortunate. The first two received budgets at slightly less than the previous year's rate—although the NSF had to fund more technologically centered research than the previous year—while the NIH received a considerable budget cut.

The NIH, like the NSF, had been given the charge to revisit its strategic planning principles. Congress insisted that science funding become more relevant. Scientists and the federal government had to do more to take science and apply it to national goals.

To assist American science in assessing its successes and goals, the House Committee on Science, Space, and Technology undertook its own strategic planning exercise on the health of American science. Chaired by George Brown (D-California), the committee called into question a handful of notions that had guided science and technology policy for decades: first, that funding individual researchers (rather than a consortium of researchers or some other configuration) was the best way to generate new scientific ideas; second, that basic research was best pursued primarily at major universities rather than some other locus; and third, that basic research is THE source "of fundamental knowledge that eventually leads to innovation, technological development, and economic growth."[20]

The House committee demanded certain things from any analysis. It wanted scientists more nearly to try to achieve consensus on what measures to support with federal funding and on how to structure efforts around that end; Congress wanted scientists to formulate some sort of science research policy. But it also demanded assessment tools. Scientists needed to establish objective means to determine if a policy was succeeding or failing and should be canceled. In addition to this straightforward framework, the committee wanted set aside a separate research pool to fund those researchers whose projects had been big successes: young, unproven but creative investigators, and managers whose grants had achieved significant payoffs for slight funding.

CLINTON AND THE PRESIDENCY OF INSPIRING TERMINOLOGY

Radical change in federal science seemed inevitable. As the election unfolded, Bush pointed to congressional inaction and interference and maintained that Congress had been the Democratic nominee Bill Clinton's surrogate; it was enacting the former Arkansas governor's policy. It had slashed the NSF and the NIH budgets, ended attempts to create high-speed rail transport, reduced investment in computers, downplayed space-age materials, and frustrated efforts to fund computer-assisted design and computer-aided manufacturing simply to placate labor unions. Clinton himself, Bush contended, sought to cut university overhead rates in excess of $3 billion yearly, a maneuver that would spell the death knell to universities as the bastion of basic research.

For his part, Clinton argued that the Bush administration failed to keep America competitive and he focused on the White House's refusal to assist industry in converting federal and other science into manufactured products. To outline his policy, Clinton unveiled a twenty-one-page plan for America's future. It began with this provocative statement: "Investing in technology is investing in America's future." In a single sentence, Clinton replaced the time-honored mantra, basic research is the wellspring to American prosperity, with a new equation where federal financing and guidance of technological innovation became the keystone to ensuring American economic success.

Clinton's paper laid out his program in what would emerge as a characteristically Clintonian way. Bombast, written as simple declarative statements, coupled with unspecified but optimistic-sounding terms, seemed to promise different programs to different people. For example, the Clinton program pledged that it would be "forging a close working partnership," "redirecting the focus," "reaffirming our commitment," and engaging "in world class research." The proposal held to "tough standards and clear vision." It would "remove obstacles," "reinvigorate," engage in "effective management," "invest in the future," and "enhance competitiveness." "Mission-oriented R&D" would yield "to investments designed . . . to strengthen

America's industrial competitiveness." A Clinton administration would establish "a world-class business environment for innovation and private sector investment."[21]

That Clinton's proposal was written in these terms was no accident. Like Bush, he had embraced Total Quality Management. In fact, Clinton had mandated as governor of Arkansas that TQM become the sole means of conducting governmental business in that state. He adapted TQM methodology to his presidential campaign. The New Age-style rhetoric resonated with policy wonks, high-tech partisans, and others who adamantly pictured a future far different than the immediate past. Clinton's use of their terminology suggested a certain kinship with those groups, that Clinton could feel their pain and because of his intimate emotional connection, that he could devise a means to set things right.

Clinton had a more substantial history of listening to and working with these groups. When in Arkansas, he had regularly attended Renaissance Weekends at Hilton Head for every New Year's Eve. There, scenarios for the future were habitually spun off by persons—academics, think-tank social scientists, Silicon Valley businessmen—explicitly seeking access to power. Almost without exception, these persons focused on refashioning government as the engine to create their improved, streamlined, postmodern programs. Listening to these New Age dreamer wonks helped shape Clinton's verbiage and probably his vision.

A careful reading of the Clinton plan identifies its parameters. He focused almost entirely on economic growth; everything in the document was justified as fostering or ensuring an expanding, technology-based economy. This expansion involved tax credits for corporations to stimulate research, experimentation and small business investment, an international open trade policy, a federal "regulatory policy that encourages innovation and achieves social objectives," and government support of "private research." It called for creation of numerous small federal contract research and development centers, a national telecommunications infrastructure, international science and technology agreements, and dual-use programs within the Defense Department to stimulate private projects. Al Gore, Clinton's vice presidential candidate and former senatorial antagonist to Bush's science policy, would oversee the entire administration's technology effort. Perhaps as a sop to constituents, the Clinton plan would reestablish "technological leadership and competitiveness of the US Automobile Industry" and invest "in energy-efficient federal buildings." The plan's goal was simple: to produce "economic growth that creates jobs and protects the environment." A Clinton-led government would transfer defense R&D money to the civilian side until both were equal. It would devote at least 10 to 20 percent of the Energy, NASA, and Defense budgets to forming R&D partnerships with industry. Because of these two initiatives, the plan projected, Congress would end earmarking so as not to undo or dilute the new federal efforts.

Of special note was Clinton's proposal to commercialize potentially rewarding products before others outside America did; the federal government, not market forces, would select the winners and losers in the new technological sweepstakes. Here as in so many other instances, Japan again seemed the villain as well as the model for the reinvention of the industry-federal partnership. The federal government would foster and fund regional technological alliances and invest in agile manufacturing networks to speed technologies into commercial products. America would be dotted with manufacturing extension centers to give industry a head start in these potential lucrative areas.

At the heart of most of these proposals was the information infrastructure. Communication—the speeding and sharing of information from place to place—seemed the solution to almost any economic issue. For example, new techniques could be diffused throughout industries quickly and at little cost. Knowledge could be transmitted from place to place almost instantaneously. Indeed, the plan went so far as to claim that through the information superhighway schools "can become high-performance workplaces."

Mentioned much less prominently and with almost no fanfare nor detail was the expansion of basic research at universities and national labs through NSF and NIH grants. Clinton's formulation accentuated economic questions to such a degree that it included basic research almost as an afterthought. Gone were the elaborate justifications and explanations of the previous forty years about basic science's sanctity and nearly mythical status to provide the information necessary to resolve any military, social, economic, or other question. In its place was a single sobering thought: the federal government needed to help industry to develop commercial goods to secure the nation's economic future.[22]

Bush greeted the Clinton plan by extending his program for Department of Commerce–industry partnerships. In October 1992 the Bush administration entered into more federal-private industrial agreements than it had during the previous four years. His belated support for an increased national government–manufacturing interface failed to draw much attention. Bush's record and apparent dependence upon science as a progenitor for the future did not galvanize scientific interest. In later October the Democratic challenger announced creation of a Council of Scientists and Engineers for Clinton/Gore—nearly seventy-five luminaries strong—to assist the ticket in scientific and technology policy matters. The Bush campaign did not respond in kind, relying instead on department and division heads to make his pro-science case.

That scientists would form the Clinton/Gore council was curious. Its chair admitted that "the Bush Administration has done rather well by science." While Clinton promised to meet with the group during the campaign, he never scheduled a meeting. More significant, his campaign staff showed no real affinity for scientists or their concerns. The campaign's full-time science and technology policy coordinator, Tom Schneider, ran a "management consulting firm that puts together leveraged buyouts of failing companies." Its other science/technology policy advisor, Ellis Mottur, had been an aide to Ted Kennedy and proposed granting public interest groups voices in otherwise peer-review professional councils. His main job during the campaign was "recruiting endorsements from business leaders." Lederman, the Clinton/Gore council vice chair, provided the group's rationale. It planned to present to Clinton policy statements on "subjects such as the environment and global overpopulation—topics that are 'not very popular' with the Bush Administration."[23]

Three possibilities account for the council's creation. First, it could have been so disturbed by the Bush administration's global warming and population positions that it was willing to sacrifice scientific support generally for backing in those two instances. Second, they could have been true patriots, willing to place the American economy in front of all else and thus sacrifice science on the mantle of technology. Third and most likely, council members were so secure in science's place within the American ethos and among the American public that they could not believe that Clinton would not do as much or more for science than had Bush. A corollary to this third position was that scientists, especially council members, would have more tangible control over where and how Clintonian science moneys were distributed and spent.

TECHNOLOGY OVERALL

Clinton's election was initially heralded in most scientific and technological quarters. But quickly reality set in. The new administration began to refer to Gore as the technology czar and promised that he would have a fundamental role in all science and technology matters. In fact, Clinton said that Gore would have "the responsibility and authority to coordinate the administration's vision for technology and lead all government agencies, including research groups, in aligning that vision." While many saw Gore as sympathetic to scientists' issues, his assent to the pinnacle of policy power disturbed them. For the first time, science policy advice would be organized by and under a layman, someone without technical or scientific training. Gore was a politician, a congressman, a member of a group held in contempt by many scientists and engineers. Clinton's statement that the president's science advisor and the White House Office of Science and Technology would have a measure of autonomy and status similar to that under Bush failed to reduce anxiety. So too did the realization that the Clinton effort so highly emphasized technology that its implementation would almost certainly hurt science. Money that had gone to science would now go to establish federally funded manufacturing centers. More ominously, scientists wondered if the NSF and the NIH budgets—the two budgets that directly supported much of American basic science—would be reassigned and redistributed to cover technological matters. In the NIH, for example, money that had gone to AIDS research might now go to AIDS treatment. Similarly, NSF funding to determine the parameters of photovoltaic cells now would go to NSF funding to establish the most effective means to manufacture them.[24]

The situation appeared even more disquieting in the first months of the new administration. Clinton's selection of John H. Gibbons as his science advisor drew considerable comment. Gibbons, a physicist involved some years earlier in nuclear waste disposal, did not come from academia and was not well known within the scientific community. Instead, he came from Congress. He had spent the previous thirteen years heading Congress's Office of Technology Assessment, the very agency that questioned the traditional science-government partnership. There he focused on environmental policy and energy conversation, working closely with Gore. Critics agreed that Gibbons's appointment sent "a strong signal" to scientists that Clinton would emphasize technology rather than basic research. Gibbons, as a "creation of Congress," would cause "a lot of paranoia" about Clinton's basic science position. It was likely, they concluded, that the new president had "fallen prey to the view that the problems of society" could be solved through "applied research."[25]

Gibbons himself contributed to the scientists' discomfort. He hired as associate directors of the White House science and technology office his three former aides from the technology assessment office. Congress itself sent similar signals. Nearly half of the fifty-five new members in Congress sought assignments to the House Committee on Science, Space and Technology. What had formerly been an exceedingly unattractive post now appeared to be critical to the nation's future. And it also seemed crucial to the new Congress members' prospects. A massive infusion of technology funding would enable savvy representatives to channel moneys to their districts to preserve jobs or to create new ones. This "very solid parochial interest" was not limited to any one part of the country.

Technology and science had become so diffuse and embedded in almost every endeavor that each new Congress member had a constituency dependent on federal science funding.

The mad dash to become a member of the aforementioned committee pointed to one of the issues that social scientists often highlighted. The new emphasis on technology in the legislature would merely replace pork barrel science with pork barrel technology. Each Congressman would be under increasing pressure to deliver technological funds to their districts and so would expand the definition of critical technologies beyond any semblance of meaning; there would be no rational discrimination. The result would be massive waste and incompetence, boondoggles of unprecedented scope.

George Brown, longtime chair of the House Committee on Science, Space and Technology, demonstrated a less than sympathetic view of scientists' concerns. Until the recent past, science policy was "designed primarily by scientists for scientists." Now society must be served. Scientists needed to understand that the federal government will only fund science "in pursuit of explicit, long-term goals." Scientists must "put their individualism at the service of larger national goals." Daniel Koshland Jr., then editor of *Science*, disagreed with the new view but recognized it as inevitable. Calling it "a monstrous policy error to cut back on basic research," he was resigned that "for the foreseeable future" Congress-defined "social goals" would be the principal grounds for "federal support of scientific research."[26]

Word came from the White House that some Big Science projects, most notably the Superconducting Super Collider and the space station, might have to be rethought. The former now had a price tag of more than $11 billion—$2 billion had been spent—while conservative estimates put the cost of the station at some $30 billion. At the same time, others were suggesting new Big Science initiatives. Launching a program of Manhattan Project proportions to cure AIDS gained considerable support.

Clinton rarely shifted focus from devising ways to assist technology. His most impressive effort revolved around the Commerce Department. He advocated spending $17 billion over four years to enhance American competitiveness. Central to this measure was transformation of the NIST into a civilian complement of the DARPA. But even that maneuver was wracked by controversy. Senators from several states housing national labs demanded that the Energy Department be vested with the primary technology-enhancing responsibility. "Don't reinvent the wheel," one commented. Energy "has scientific and technical capabilities and resources within the departmental laboratories in virtually every area of importance to the economic, scientific and technological competitiveness of United States industry." We, he concluded, "see no reason to set up new programs."[27]

As Congress and scientists tried to find their way within this new Clintonian technology-centric framework, they were thrown an additional curve. Within two months of his inauguration, Clinton proclaimed the "end of big government" and created the National Performance Review. The NPR's focus was not on a specific program or policy or thing, or even governmental performance. Its objective was to increase satisfaction within the federal government itself. Clinton had expressed concern before assuming office that "confidence in government," defined as "simply confidence in our own ability to solve problems by working together," had "been plummeting for three decades." He recognized only two options: "Either . . . rebuild that faith or abandon the future to chaos." Clinton wanted "to change the culture of our national bureaucracy away from complacency and entitlement toward initiative and empowerment. We intend to redesign, to reinvent, to reinvigorate the entire National Government."

Clinton placed Gore at NPR's helm. "Corporate America, [which] had reinvented itself to compete and win" the 1980s economic fracas with Japan served as Gore's reputation-resurrecting exemplar. He understood that those battles had not been fought in factories but in management seminars and so he created an army of 250 civil servants, trained in or experienced with these new management principles and techniques. Gore charged the group to focus on how government works, rather than on what it should be doing, and to concentrate not on "moving boxes" within the organization chart but "on fixing what was inside the boxes." The management-savvy review team drew almost exclusively from the management theory with which its members were acquainted—Total Quality Management—and quickly coalesced around "a clear set of principles and an inspiring vision of what government should look like." Their epiphany resulted in the mantra: they "would create a government that works better and costs less by putting customers first, empowering employees to allow them to put customers first, cutting the red tape that held back employees, and cutting back to basics." Gore then gave each agency until early September to study itself and propose action, forbidding as "not acceptable" recommendations for "further studies."[28]

Clinton's bold National Performance Review fit the persona he was trying to create. He claimed to be a "New Democrat," fiscally responsible and socially active, not wedded to the past. He projected dynamism, confident positive action. He exuded command. His agenda was all about reinventing, making the present a bridge to the twenty-first century rather than a hollow reminder of the twentieth; he was less Roosevelt than he was New Age. He called himself progressive and rejected attempts to portray him as liberal, which connoted an older failed approach. In this view, Clinton would never trumpet a national industrial policy. Such a creature was the product of dated thinking. Dated thinking in the Clintonian vernacular was thinking in which there already were protagonists and antagonists, where people had taken hardened positions. He would push technology policy.

Clinton offered style as much as substance. How things were cast and phrased, and what the public saw and heard were central to his ability to govern. Such was the raison d'être of the National Performance Review. Clinton used social science extensively to hone his image and hence his ability to govern. (It was no surprise that the Clinton administration favored increased social science funding. Much of the new funding went not to competitive grants but rather to contracts. There, terms could be specified in much greater detail and peer-review requirements cast aside). Focus groups and polling were regularly employed. No question was too insignificant. Clinton even used focus groups to determine at which venue to take a vacation. (See fig. 10.)

http://www.nyc.gov/html/dep/gif/hammer.jpg

Fig. 10

Department of Environmental Protection employees Larry Beckhardt and Ed Blouin receive the coveted "Hammer Award" from Vice President Gore's National Partnership for Reinventing Government for their work on the Conservation Reserve Enhancement Program in the New York City watershed in 1998.

At the heart of the Clintonian agenda was a real fear that the consensus on government as a positive force in American society had declined precipitously. Americans lost faith in the ability of government to resolve problems. Science as a means to evaluate and effect change also was viewed with a new sense of suspicion. Clinton perceived these declines as emotional, not rational. His thrust was to modify the image of government. He remained convinced that it did good work. What he needed to do was to restore its face and by extension its place within America. Through image building—smiling employees, increasing access, reduced paperwork, and the like—Clinton expected Americans to again embrace government's positive role in American life.

Clinton then adopted a management philosophy different than had Bush. Bush managed stakeholders and the various facets of government. Clinton managed the ability to govern. He knew what he wanted to do and where to do it. His initial thrust was to generate support for government and for his programs, and that led him to concentrate on ways to present them; he advertised himself and marketed the goals he hoped to achieve. A critical method to accomplish the latter was the manipulation of image, vocabulary, and details. No matter was too small to be ignored.

A serious problem would emerge from this image fest. It was not always easy to understand where Clinton's marketing agenda ended and where his "true" principles began. If the ability to govern was conditioned on image and nuance, then anything that undercut image hampered governance. To have hopes of resurrecting government mandated that image be restored or repaired. And that reparation/restoration had to take precedence over a particular program or plan. Very simply, Clinton often had to abandon positions or sacrifice people he had backed when they became a liability to the image he had hoped to present. Similarly, Clinton found himself embracing positions and people primarily because of their ability to foster a particular image.

During the first six months of his presidency, Clinton promised to reinvent the automobile as a green, environmentally friendly device. He also argued that environmental technologies were growth industries; rather than costing money, environmental technologies could be developed and marketed as critical profit centers. He pledged to reinvent the mission of national laboratories, especially those long involved in weapons manufacture, and to have them serve American industry. He reinvented basic science as "curiosity-driven research" and applied science or technology as "strategically targeted research." He claimed to be working to reinvent the economy and was reinventing healthcare. His administration had several meetings with Silicon Valley entrepreneurs and others representing newer economic possibilities. He appointed numerous women and minorities to significant places within his administration. He urged funding for women's health issues, breast cancer, and AIDS at unprecedented levels. He established within the NSF and elsewhere a program of what his detractors called "presidential earmarks" where he mandated that a precise percentage of the agency's funding go for "strategic research" into "critical technologies and national needs"—

environmental technologies, advanced computing, biotechnology, materials processing, and the information superhighway. Indeed, the Clinton administration claimed that government "will no longer rely on 'serendipity' to produce commercial technology," but would rather aggressively pursue industry-government partnerships to create profitable consumer and other products.[29]

In a rather perverse way, Clinton's NIH and NSF budgets mimicked programs in his predecessors' administrations. There, Big Science took big money, leaving significantly less for other matters than science proponents wished. Under Clinton, special interest money— for breast cancer, AIDS, women and minority health (Clinton himself argued that when it came to healthcare, women "were treated like second class citizens"), super computing, and the Human Genome Project—captured virtually all the new funding that the administration proposed for those agencies. Other kinds of science frequently saw a contraction of their budgets even before inflation.[30]

This budget strategy irked scientists in those stagnantly funded disciplines. Daniel Koshland, the editor of *Science*, sniffed that Clinton's failure to prioritize basic science was opting for "jobs now versus jobs in the future." Others argued that Clinton's healthcare plan would kill the American pharmaceutical industry and doom biotech research. Social scientists in particular argued that not expanding biotech and biomedical research now would inevitably produce much higher health costs later.[31]

Clinton's decision to seek a smaller, less costly space station, coupled with Congress's willingness and the administration's assent to killing the Super Collider, freed up some funding. In hopes of making America "the green giant" of the world, Clinton chose to support new environmental technologies. Establishment of a National Environmental Technology Agency to serve as a clearinghouse for environmental design figured prominently in his thinking. There, ways to reduce global warming, contain nuclear waste, and control pesticides could be manufactured. But that technology-based framework disappointed scientists who countered with a National Institute for the Environment. Arguing that environmental research in America had been about "putting out fires, and cleaning up train wrecks," they proposed research as a remedy for environmental disaster. Research amounted to prevention, they argued, and suggested a wide-ranging research agenda. Among the things they recommended was a $141 million program to examine how ultraviolet radiation can "ripple through the food chain" as well as its effects on marine organisms. Forty Congress members lined up to support this plan. The NAS offered its own alternative. It found "no comprehensive plan of environmental research" that coordinated research already taking place. Lack of "clear leadership" in environmental research hampered efforts and the academy recommended creation of a National Environmental Council in the White House. Chaired by Gore, the council could systematically identify environmental questions and fund research for their resolution.[32]

Beyond this organizational battle rested another factor: regulation. Critics claimed that environmental regulation was capricious or political. It either supported an ideology or was governed by hunches. Partisans demanded regulation based upon the best science. That demand stemmed from two basic assumptions. First, that best science was the best available knowledge and therefore the most accurate. The second assumption suggested that the best science was the cheapest; it reduced waste by requiring nothing to be outlawed that was unnecessary and protected the public within the bounds of reason.

The Clinton administration found it difficult to find a high-profile environmental scientist to head the EPA's scientific effort. No one wanted to lead a division in which the objectives were impossible. Sound environmental regulation required many-year, multi-layered research. But the EPA was called on regularly to respond to some situation, catastrophe, or event; regulation was a political as well as a scientific determination. Money, staff, and equipment were yanked from one study and reassigned as a new mission was put forward. EPA's ethical guidelines were even more constricting. Because of the political sensitivity of environmental regulation, the agency's general counsel placed a five-year moratorium on seeking to use influence or work with the agency for any major official leaving EPA service. In practice, this meant that a science director when he stepped down would not be eligible to apply for EPA grants for five years and thus would have to abandon or sharply reduce his/her scientific research for half a decade, several lifetimes in research terms.

The quagmire over science at the EPA and the dispute over how to organize environmental research were illustrative. Political divisions among branches of government and interest groups proved more powerful than a desire to systematize environmental research. Indeed, forming a constituency to back a particular type of investigation was no guarantee that the work would gain political favor or that bureaucratic matters would not undercut the activity. Yet it would be incorrect to suggest that the Clintonian blitzkrieg failed to move government. In the latter half of his first year, Clinton had established a consortium linking the Big Three automakers and government to produce green automobiles to triple MPG ratings; agitated for new pesticide legislation, which would ban the most toxic chemicals while permitting use of those with weak carcinogenic properties; and redesigned the research agenda of Los Alamos and Lawrence Livermore, two large federal laboratories that had specialized in nuclear weaponry. NSF social scientists issued studies showing that African Americans, Hispanics, and women all were making steady progress in math and science achievement, that women were still not joining the ranks of scientists in anywhere near the percentage of their population, and that racism hampered the careers of black scientists. The NIH unveiled a big new program, organized around a major conference, to make research into disease prevention a major priority. Rather than attempt to heal or ward off disease, this new research agenda focused instead on what constituted health and what made humans liable to diseases.

Momentum generated in the reinventing government initiative also found expression in science policy. Gore proposed a National Science and Technology Council similar in scope and power to the National Security Council to formulate US science and technology policy. As Gibbons defined it, the council's benefits were chiefly bureaucratic. In addition to giving "technology policy a higher profile," the council would "enable the White House to referee disputes over regulations and budgets among federal departments." It gives them, Gibbons concluded about the newly approved council, "more clout to knock heads between various agencies."[33]

While the White House offered one means to streamline science and technology administration, a bipartisan group of congressmen took the matter further. They hoped to create a single department of science by folding together the Commerce and Energy Departments, NASA, the EPA, and the NSF as well as the White House science and technology office. That consolidation would save an estimated $100 million just in administrative costs. Coupled with plans to limit overhead rates to 50 percent, the plan's proponents maintained it would reduce expenses by more than a half billion annually.

THE SCIENCE PRESIDENT

While Congress granted the White House no new organizational authority, it did agree with Clinton about changes that needed to be made in the General Agreement on Trades and Tariffs. The initial draft of the GATT, drawn up under Bush and aimed primarily at the Japanese, permitted nations to levy substantial tariffs on goods derived from government-industry partnerships. But with Clinton's forceful moves to create such partnerships, that section of the measure now proved awkward. Congress petitioned to get US trade negotiators to remove the stipulation.

That effort proved successful. Roadblocks failed to stop the reinvention express. Under the Office of Science and Technology Policy, nearly three hundred scientists and social scientists involved in science policy formation were invited to Washington to help define the administration's position on "fundamental science." Such a meeting was necessary, the office argued, to ward off "a growing public feeling that research is a luxury the country can no longer afford." A "new national strategy for science" can become "a rallying cry for researchers to explain what they do and why it deserves funding." The White House's ability to get its science/technology agenda through Congress did little to reduce anxiety. Although Clinton got almost everything he asked for in his science/technology package, margins were uncomfortably slim. "Cut-cruisers" and "budget hawks," congressmen whose "obsession [was] balancing the budget," appeared in the ascendancy.[34]

The Clinton administration recognized the narrow passage of its science/technology legislation as a major obstacle to its long-term ability to govern. Science and technology had become so ingrained in virtually every aspect of governance that a precipitous decline in funding or even flat budgeting undercut the ability to govern. Regulation, the economy, health, safety, and other facets too numerous to mention now seemed to depend directly on scientific and technological action. Clinton attacked the problem of what to do about science in the way he attacked other problems: by attempting to create a positive image and thus gain critical political support. He would concoct and issue a vision statement.

The fundamental science meeting had been the first maneuver in that image-building process. Its culmination was publication of a major document, *Science in the National Interest*. Hyped as the first definitive science policy statement since Vannevar Bush's *Science: The Endless Frontier*, the Clinton document began by grandiosely proclaiming "science: the endless resource" and then considering the traditional basis of science funding in America. It argued that the cold war had yielded a defense-related science policy and that that policy's success had led to America's world preeminence. The Soviet Union's collapse removed the specter of communist domination and necessitated a new justification and policy. Global economic competition, sometimes aided by national governments, emerged as the primary explanation for massive scientific investment. Without a sound economy, America would be lost. Indeed, Clinton argued that these new economic challenges posed a threat to America no less menacing or immediate than had world communism. In order for America to maintain global leadership, it needed to be the predominant economic power, which required scientific and technological prowess. Only through scientific research "could we enlist the forces of the natural world to solve many of the uniquely human problems we face—feeding and providing energy to a growing population, improving human health, taking responsibility for protecting the environment and the global ecosystem, and ensuring our own nation's security."

A long exegesis culminated in five main national science policy goals. Science policy must: "maintain leadership across the frontiers of scientific knowledge"; "enhance connections between fundamental research and national goals"; "stimulate partnerships that promote investments in fundamental science and engineering and effective use of physical, human, and financial resources"; "produce the finest scientists and engineers for the twenty-first century"; and "raise the scientific and technological literacy of all Americans." Much of the remainder of the document was a consideration and demonstration of how then-current Clinton administration policies would achieve those ends. In addition to self-serving rhetoric, the policy brief called for a more rapid conversion from defense to civilian science as well as heavier investment in science.[35]

Even critics hailed *Science in the National Interest* for its eloquence. Scientists of all stripes generally proclaimed the document a success, if for no other reason than for raising the scientific community's morale. But Clinton did little more than offer the report to bolster scientific spirit. He remained notoriously remiss in consulting or even convening the science policy groups he had created. The National Science and Technology Council, established in November 1993, did not hold its initial meeting until seven months later. The gathering lasted a scant thirty minutes. Similarly, Clinton's external science policy group had its first meeting almost a year after it was formed.

Clinton clearly did not depend on either of these groups for advice when he and Gore issued their manifesto. But it would have apparently done little to sway congressional opinion. Even congressmen long known for their support of science termed the plea for more science money as "highly unrealistic." Congress's science budget for 1995 proved that the case. Its science and technology funding was flat but the considerable mandated areas—strategic research sectors—meant that virtually all other facets suffered considerable reductions. Yet that was not the least of the Clinton administration's difficulties. An unprecedented (and to scientists an unexpected) midterm electoral landslide gave Republicans majorities in both the House and Senate. For the first time in forty years, Republicans controlled both congressional houses.[36]

A "Contract with America" had fueled the Republican tidal wave. Among its most prominent planks were pledges not to raise taxes, to balance the budget, and to reduce the size and scope of the federal government. Zero-based budgeting and a congressional line item veto were its primary tools. "Everything should be looked at from the ground up," demanded Newt Gingrich (R-Georgia), the contract's architect in the House. White House operatives predicted a "scorched-earth policy" in the new Congress. Others were a bit less pessimistic. They recognized that Republicans often supported initiatives like the information superhighway but despaired that attempts to establish private-federal industrial partnerships would almost certainly be doomed.[37]

Fifty-three percent of the Republicans in the next Congress would have two or fewer years of experience. This unprecedented turnover, coupled with the contract's terms, meant there would be no business as usual. Agreements, procedures, and methods long held sacrosanct now held little credence, manifestations of failed policies. Not only would every committee have a new chair, but rules for committee action and protocol would also be redrawn. "Science and technology policy headed for political maelstrom," yelled *Science*. But cooler heads remembered that the Reagan-Bush years had been good for science. It was industry-federal partnerships—technology policy—that were rejected. Would money that under Clinton went to technology policy now go to science or would the whole enterprise be severely reduced?[38]

The Clinton administration took the lead in declaring that any cuts in its programs likely would damage American industry for decades to come. Interior Secretary Bruce Babbitt, for instance, responded to Republican suggestions that the US Geological Survey and the Bureau of Mines be closed as tantamount to "book burning" and "burning a few more heretics at the stake." He carried his complaint to the AAAS. "Science is what made this country great; it is what has carried it so far," he maintained. Babbitt further argued that science in the Interior Department was merely a Republican scapegoat. They, he claimed, really wanted to punish his department for enforcing the Endangered Species Act and the Clean Water Act.[39]

While Republicans mentioned the Interior as a place of scientific cost cutting, the Clinton administration selected Defense. There, nearly eighty labs had made the Defense Department the government's largest R&D enterprise. But the cold war was long passed. White House science advisors thought economies could be achieved by closing and consolidating as many as twenty-four labs. The method they proposed mimicked the means that the government closed military bases. An independent commission would recommend which to close or merge. The president then would approve the package, requiring Congress to accept or reject the list in toto.

When confronted with the list of labs scheduled for closure, Republicans in the new Congress approved it. They then focused on regulation and authored the "Risk Assessment and Cost-Benefit Act of 1995." It required each possible regulatory act to be accompanied by both risk assessments and cost-benefit analyses. Clintonians objected to this approach. They complained that it was relatively easy to discern economic costs but difficult to anticipate or measure benefits or risks. Their critique ignored the fact that risk assessment/cost-benefit analyses had been the basis of most federal regulatory activity for over two decades. Getting the deficit under control was by far the Republicans' most prominent effort. Several bills introduced in Congress promised to eliminate the deficit by 2002. Estimates varied but deficit eradication required slashing the federal budget by between $1 and $1.4 trillion over the next five years. Proposals included chopping a billion dollars from the NIH during that period, ending the NIST's industrial research programs, cutting NASA's earth-monitoring research, terminating some energy research initiatives, and stopping the NSF's support for social, behavioral, and economic science.

Science was appalled by these proposed cuts, even as it noted that most programs outside of science and technology received much, much harsher treatment. With typical scientific arrogance, many of the journal's constituents explained that the Republicans obviously did not understand science, for if they had, surely no cuts would have even been suggested. Congressional Republicans argued that science had gotten off lightly and that it had been their long-stated plan "to go after applied R&D," which they sometimes called "corporate welfare." Congressional Democrats claimed that Republicans took "a meat ax approach that puts our nation's future at grave risk." Their treatment of science "represents a retreat from the federal government's historical role as a driver in research and development."[40]

In what became a characteristic ploy, Democrats repeatedly used the term science to refer to both science *and* technology. After years of discussing technology policy, the phrase tended to disappear. The idea did not. Democrats found it more fruitful to use only the term science because it had both a much broader following and it carried a sense of mystery and awe. Americans backed government science. Only some supported government technology. By calling everything science provided the benefits but few of the drawbacks. Using only the term science also enabled Democrats to mobilize the "scientific community," a relatively vocal group. Its sense of solidarity would lead it to respond to any threat to science as a menace to scientists generally.

The Office of Science and Technology Policy initially received much blame for proposed scientific budget cuts. Brown complained that the office was not producing. A proposed 3 percent across-the-board cut that included the NIH and the NSF led to further criticism. Some scientists condemned the Clinton administration for issuing *Science in the National Interest*. It unwisely raised hopes and morale, leaving them despondent when Republicans took charge. Still others criticized Gore personally. His interest in the environment and passion for reinventing government caused him to ignore science. The result was widespread devastation of the American scientific community. Crunching its own numbers, the AAAS predicted that the Republican plan amounted to a decline by one-third of American scientific research funding within seven years.

By fall 1995 Democratic legislators began to urge the Clinton administration to defend its technology programs directly. Clinton should emphasize job creation, high-paying jobs. "It's good policy to promote growth and high-wage jobs," reminded a congressional staffer. The Republicans' "misguided approach to a balanced budget" is "cutting the source of our economic growth." When the "federal government reduces R&D investment . . . the private sector does too." Standing in for the president, Gore came out swinging in late October. Republican cuts were "unwarranted, unwise, and unnecessary," an attack on the entire R&D enterprise. "We are philosophically opposed to people cutting out whole areas of research," added a Gore aide. Jay Rockefeller, (D-West Virginia) injected his own rhetorical flourish. If the Republican R&D budget were adopted, "the next generation of scientists and engineers will be Japanese and Germans, not Americans." Robert Walker (R-Pennsylvania), chair of the House science committee, demurred. He maintained that the administration was merely "defending the way things have always been, and . . . has no designs to accommodate science programs to the changing world." Republicans "are looking 20 to 25 years in the future." The administration is "looking backward."[41]

The Clinton administration's decision to pummel Republicans with the same vehemence it thought it had encountered marked a change in the administration's management philosophy. While it continued to present itself as progressive and its policies as new and enlightened, it increasingly looked inward, toward its political base. Faced with the problem of a contrary Congress, the administration concentrated on suggesting to its traditional constituencies that their futures, no matter how meager and blighted, depended on Clinton. He kept the Republican wolf from the door. No matter how terrible a situation became, the Clinton administration argued it would have been far worse had it not existed. Put baldly, Clintonians claimed that the only hope for traditional Democratic constituencies rested in the administration.

That sensibility further discouraged compromise. Anger, frustration, and intense disillusionment were its staples. So too were extreme positions and fervent loyalty. To maintain that one's success rested entirely on the existence of the administration meant that the administration needed to demarcate its position as radically different from an alternative; not to do so would undercut the "need" for a counterweight. But by restricting themselves to a particular party or philosophy, a group eliminated whatever political power it might have had. It could not play one faction against another. The result usually was that the party to which it was beholden tended to take it for granted.

The science and technology debate took place within a broader controversy over the deficit. For three weeks—December 15, 1995, to January 7, 1996—the federal government closed as the Democratic president and the Republican Congress stood deadlocked. Whether a matter of principle or to make political hay, both sides postured over the 1996 budget, which was supposed to have taken effect on October 1. But the reopening of government did not mean that a compromise had been reached. The parties agreed upon a framework for deficit reduction but the entire matter continued for several more months. A series of temporary budgets—some as short as two weeks—kept government operating. Scientists were horrified that the final agreement might freeze them out. It wasn't that Congress and the president may have been in an "anti-science mood," worried the American Physical Society's Robert Park. "It's that we are not even being considered." NSF director Neal Lane urged scientists "to work together in ways we have never done before, to raise our voices, together." Others were more pessimistic about scientists' ability to change their fate. In this budget battle, which focused on Medicare, Medicaid, and the deficit, science funding was like "water drops to a shaking dog." Science was so insignificant in the White House, they argued, that Clinton even failed to mention it in his yearly State of the Union address.[42]

With the framework for deficit reduction in place, Gore ratcheted up the partisan heat. The Republicans were "approaching science with the wisdom of a potted plant." Congress was advocating "a retreat from understanding, a know-nothing society." Its science policies were "appropriate for Fred Flintstone." He reminded his audience that the administration supported science. Its proposals included "generous amounts for science and technology programs in comparison to Republican proposals."[43]

When the final budget compromise was announced, science fared well. The NIH got a nearly 6 percent boost from the previous year and the NSF also received additional funding. Overall, civilian basic research received a 3 percent bump from Fiscal Year 1995. Programs for industrial-federal cooperation in both the Departments of Commerce and Energy took the largest hit. A similar distribution occurred in the budget plans for the next fiscal year, which ironically were issued at almost the precise time of the budget compromise for 1996. Republicans offered a 5 percent basic research jump but took the money from the Commerce Department's industrial partnership funds. Clinton's budget offered less for NSF and NIH basic research but included an additional $1 billion for private-public research. "There's really been no change in our priorities," the administration maintained.[44]

Nobel laureates may have concurred but they certainly did not like the direction of Clinton's program. In a curt letter delivered on June 19, 1996, they complained that the president wasted resources and political capital fighting for industry-federal technological interplay. At best, they continued, federal assistance to technological ends was a subordinate issue and they urged the president to "reaffirm the fundamental role of the federal government in supporting basic scientific research." That was not the only bad press to hit Clinton that summer. Walker called the heads of NASA, the NSF, and the Energy Department's research program to explain why they had criticized the Republican-proposed science budgets for their agencies for the next four years yet loudly supported Clinton's budget recommendations for the same period, which in each case called for significantly greater reductions. Administrator after administrator replied that Clinton budget figures were accurate with respect to gross spending on science, but each promised "one hell of a fight" to increase spending in his area. They in effect argued that the Clinton figures presented to Congress were simply the basis for discussion, not final totals, and that they were confident that the president would rearrange his priorities. Walker accused them of having it both ways. On one hand, they could talk of the deep cuts planned by Clinton and thus outsaving the Republicans, while on the other maintain that Clinton was a friend and fervent supporter of science.[45]

The noted science journalist Daniel C. Greenberg went further. In a withering column, Greenberg pronounced Clinton the master of "political chutzpah" for pushing his science record. Mentioning on every occasion "computers, the Internet, government grants that led to Nobel Prizes, survival of government laboratories on the Contract with America hit list" and much more, the president's "head-spinning torrents of . . . campaign rhetoric . . . create the impression that a patron of scientific inquiry inhabits the White House." The reality was quite different. Although Clinton had taken credit for science spending, it had been Congress that authorized science budgets far in excess of what Clinton requested. "Clinton has paid little attention to fundamental research, the intellectual and underfinanced core of the scientific enterprise." There are no "scientists and engineers in the president's inner circle"; "throughout his adult life he has made no connections in the scientific community." Instead, Greenberg countered, he has wasted "a good deal of political capital and government money" by "expanding the federal role in industrial research." Such an approach was preposterous. Industry already invested more than $100 billion of its own money, three times greater than the annual federal science budget. Finally, Greenberg reminded readers that industry had been "cool to federal research assistance"—it did not want government funding—arguing that tax and regulatory relief would prove much more beneficial.[46]

Congress had become at least as powerful as the president in science and technology policy formulation. The executive branch proposed rules for embryo research and Congress tacked on prohibitions against certain practices. Clinton centralized NASA authority and provided it with the responsibility to do things cheaper, faster, and better—reusable spacecraft, unmanned missions, approachable spokespersons. The feed from NASA's voyages was sent in real time over the Internet for geeks worldwide to enjoy. But Congress also made itself felt. The agency's ambitious Project Earth was scaled back at congressional insistence and the agency marketed itself as being a science-driven entity. By executive order, Clinton appointed a board to consider ethical issues arising from human experimentation but it lacked power beyond holding hearings. Congress also took up the environment. It favored and funded an inventory of every species of animal and plant in the United States. It seconded

Clinton initiatives to make regulatory decisions in the Interior and within the EPA based on "science." Congress dropped its opposition to funding social and behavioral science through the NSF as well as the US Geological Survey.

Both the Republican Congress and the Democratic White House staked out very different positions, argued tenaciously and bitterly, and claimed the other's policies were going to ruin America. Yet they managed to hammer out compromise after compromise. That was no accident. Underneath the antagonism, contempt, and poisonous rhetoric rested a sense that America's future depended on science and technology and that the federal government had a critical responsibility to fund and direct those endeavors. They disputed the parameters, of course, but sites of contentiousness were far less significant for policy formation and implementation than the broad areas of agreement. Rancor continued unabated solely because conflict, partisanship, and virulence worked to partisan advantage. It cast aspersions on the opposition, potentially weakening it, while compelling supporters to cleave ever more closely. It was uncivil but effective. The new paradigm for electoral success—now almost the only thing that mattered—was less about capturing the political middle—a dominant electoral technique for much of the century—than isolating one's opponent in the corner. Painting the opposition as a dangerous radical force on the fringe of the American political scene emerged as the dominant political style.

Science and technology policy did not figure prominently in the 1996 election. Other issues proved more dramatic, more useful as political tools for separation and of identification. Science and technology lacked the power to galvanize broad portions of the American population, at least in part because both political parties agreed in principle over its importance. That fact was acknowledged immediately after the election. With the ballots freshly counted, there remained no imminent need to portray the opposition as irredeemably out of touch. There would be plenty of time to do that before the next election. In science and technology—areas where a fundamental agreement existed, the two parties called a truce and pledged to work together to devise an appropriate policy.

That policy incorporated a relatively heavy dose of basic science money and somewhat less funding for establishing corporate-federal technological partnerships. The NIH, the NSF, and basic science in the Department of Energy were scheduled to increase in the Clinton budget but Congress went further and provided a 5 percent boost. Clinton's budget gave generously for private-federal initiatives and Congress lowered the total some. Still, even during these lean budget times, Clinton's pet technological efforts received funding from the Republican Congress in excess of the previous year's total. Special interest earmarks within the various agencies did very well in both venues. AIDS, breast and prostate cancer, the Human Genome Project, and other projects with politically identifiable and viable communities received generous support from both the president and Congress.

Criticism of government science and technology policies as inadequate did not cease with these more munificent budgets. Several scientific organizations offered various doomsday scenarios and demanded additional appropriations. The administration and Congress took what scientists said with a grain of salt; they dismissed scientific lobbyists as insatiable, greedy, unrealistic. For their part, disgruntled scientists sometimes targeted Gibbons's performance as an explanation for their dissatisfaction. His inability to shape and meld the Office of Science and Technology Policy into an effective lobbying agent for science and technology funding in the White House translated into less money for scientific and technological endeavors. That the office often issued statements that championed congressional-executive compromises increased their disillusionment.

Perhaps because of the complaints, Gibbons made known his desire to leave the administration. As it considered possible successors, the White House rarely issued science policy statements. The Kyoto Protocol was a notable exception. To build public support for the idea, Clinton and Gore called seven environmental scientists to the Oval Office for a public seminar on the potential dangers of unabated greenhouse gas emissions. The administration knew that it faced an uphill battle. By a vote of 95–0, the Senate urged the president not to sign a protocol if it would cause the economy "serious harm" or if developing nations were not held to the same tough standards as the United States. The Clinton administration clearly had a nearly impossible task. The nation had already failed to meet a gentlemen's agreement to reduce carbon emissions that had been initialed in 1992. Creation of an Intergovernmental Panel on Climate Change (IPCC) as a tool to identify and market administration environmental pollution policy marked Clinton's next step. Its job was to hammer out consensus among the various governmental stakeholders and their supporters outside government.

The panel divided itself into a number of subpanels, each of which deliberated for a scant two weeks to speed their findings to the public. The subpanel charged with detailing the consequences of global warming presented the most dramatic report. Labeling its report "scenarios and projections," not "predictions," the group claimed that island nations would be inundated by oceans as polar ice caps melted. More severe weather would become commonplace and whole forests would disappear. Malarial disease would jump nearly 40 percent worldwide. But all was not negative. While China would grow less rice, its climate would be more suitable to wheat. America would have less expensive winter heating bills and lower snow removal costs. Scandinavia could grow winter crops.[47]

Clinton's gathered this and other information to fashion his Kyoto proposal. Delivered in Japan in December 1997, it promised to decrease carbon dioxide emissions in the United States to 1990 levels within fifteen years. Its primary mechanism was spending $5 billion over five years for research, tax credits, and other incentives for companies to reduce emissions. America, the Democratic Clinton asserted, will lead the world "to reduce greenhouse gas emissions" in a characteristically American way, "through market forces, new technologies, energy efficiency."[48]

Science, Technology, and Electoral Politics: A Bridge to the Twenty-first Century

As the Japan-based group deliberated, Clinton turned his attention to the 1998 congressional election, now less than a year away.

Clinton rediscovered the twenty-first century and chose to make that the focus of his campaign to elect a Democratic Congress. He (and Congress) had been quite fortunate. An unprecedented boom had increased American productivity and wages to such a level that the monstrously large deficit almost disappeared. Congress and the president agreed to keep a lid on spending, however. For his part, Clinton demanded that any surplus not be used for new programs but to shore up social security, the safety net for millions of Americans. Congress wanted to reduce taxes.

Clinton hailed the new millennium in his 1998 State of the Union address. Terming it an information age and an education age and unveiling a White House Millennium Program as well as a Next Generation Internet, Clinton took every opportunity to market himself and his party as cutting edge. His "21st Century Research Fund" was in the same vein. In this case, Clinton proposed something quite different than he had previously. He proposed the largest basic science increases in the nation's history; NSF, NIH, USDA, and DOE would all receive unprecedented research appropriations, as much as a 10 percent hike. Even defense basic research would grow by nearly 7 percent in the next fiscal year. Rather than concentrate on technology and especially on trying to fashion a public-private partnership, Clinton abandoned that approach almost entirely and concentrated instead on basic research.[49]

For the first time in decades, Congress and the White House agreed in principle on science policy. The administration spared no expense in congratulating itself. Clinton himself traveled to Los Alamos, the University of Illinois, and the 150th anniversary of the AAAS in Philadelphia to spread the good news. Gore conducted a series of press conferences in Washington and California to outline possibilities. The budget is now in balance, Gore assured his audience, and this new spending "reflects the nation's priorities . . . in science and technology." Harold Varmus, NIH head, maintained that "this is a budget in which everyone wins." Neal Lane, soon to be appointed OSTP chief, mused that just "two years ago, we were worried about major cuts." Now "we celebrate recognition of the critical role of science and technology."[50]

The devil was in the details. Unlike no other president, almost all the new money Clinton proposed was earmarked for particular programs. To be sure, presidents had long sought money for various causes and to placate avid constituencies. But Clinton raised that sort of constituent service and micromanagement to new heights. More critically, however, was that Clinton knew that the money to fund his proposal was going to back the Republicans into a corner. He "funded" his grand basic science initiatives through possible congressional approval of a settlement between tobacco manufacturers and state attorney generals. That agreement was to bring an estimated $65 billion into the federal coffers over five years and would place tobacco as an addictive drug within the FDA where it would be tightly controlled. But Clinton knew no such agreement was likely to occur and he did not even offer the outlines of one; the Republican Congress disliked the measure, which provided funding to dissuade future lawsuits and to settle all Medicare or Medicaid claims against the manufacturers. Many of Clinton's supporters also opposed the settlement as far too lenient for the tobacco men. People on both sides of the aisle were disgusted by new revelations of the tobacco companies' extensive, high-powered campaigns to get children to smoke.

Clinton pushed his unfunded basic science proposal as if it were a fait accompli. If Congress did not pass the tobacco deal, would the whole agreement go up in smoke? The Republican Congress could divert a portion of the projected surplus to cover the new initiative rather than apportion it to social security, but then it ran the risk of acting as it claimed the Democrats had, as a tax-and-spend body. It would also seem as if it did not care to prop up social security. If on the other hand the surplus did not materialize and with the budget agreement fully in place, Congress would have to cut other programs to raise the requisite funds. That tack would surely enrage whoever anticipated receiving those funds.

The tobacco settlement did not materialize that summer. Congress left to its own devices opted to go forward with the heart of Clinton's proposals anyway. "I'm thrilled," chirped Gingrich. "I want to see as high an increase as possible." But to balance the budget required them to cut spending and among the things the Republicans chose to reduce was a summer job program for inner-city youth and energy subsidies for the poor. Several Democratic congressmen denounced the move and Clinton threatened a veto. To guard against retaliation, Clinton refused to send his science advisor nominees to Capitol Hill for confirmation during the electoral season. Too many opportunities would exist for the Republicans to make hay.[51]

The White House kept up the pressure with symbol after symbol as the administration refused to miss any potential marketing ploy. It held what was in effect a science day and brought Stephen Hawking to deliver a lecture. (All did not go precisely as planned. Hawking publicly chided Clinton for canceling the Super Collider.) A Clinton-appointed advisory panel released a study asserting that one-third of the United States' economic growth since 1992 stemmed from the Internet and other computer-related businesses. Through executive order, Clinton created a food safety council to develop a "comprehensive strategic plan" for all science-based inspection systems and other food issues.[52]

Gingrich responded in kind. On September 24, he circulated a document titled *Unlocking Our Future: Toward a New National Science Policy. A Report to Congress.* Gingrich had commissioned this study some months earlier to ensure that the Republican Congress received credit for fostering science and technology. *Unlocking* read like a Republican primer but more folksy. Each section started with scripture or a famous quotation. Its points were several. Understanding-driven, mission-driven, and problem-driven investigations were all valid and all required government support. Virtually every activity, not just the economy or defense, depended on American science and technology prominence. Government needed to create channels through which government-financed science could become actualized as technology. Present-day science funding was barely adequate to maintain world leadership. Americans should join together to make science and technology a higher priority. Science and technology required a more prominent place within the American public school curriculum.

With the election approaching Congress had to make its science budget choice. It opted for fund increases similar to those proposed by Clinton some months earlier as a means to blunt the question as a campaign issue. To finance this new spending, Congress took part of the surplus as an "emergency" measure, which freed it from the constraints of the budget agreement. Technically, Congress did not inaugurate new spending. It merely catered to an emergency.

After the election science and technology descended into the background, except for issues of morality. Human cloning bothered Clinton and creation of a part-human, part-cow embryonic mass sent him to the ethicists. He directed his panel to review all stem cell research but also noted that stem cells might hold promise for treating a variety of diseases now incurable. What he wanted was a "thorough review, balancing all ethical and medical considerations."[53]

In effect, Clinton requested a cost-benefit analysis in which costs were moral and benefits tangible. Those were hardly easy comparisons, even for highly trained ethicists. As critical, those judgments almost certainly did not reside in the field of ethics but rather in the culture of politics. One biological group proposed to grow stem cells with fatal flaws so that there would be no chance the mass could grow into a human. In their minds, since the mass was not a potential human it was not human. But ethicists worried that by intentionally damaging stem cells so that they would not become human, scientists were in fact playing God by generating lethal mutations. Others talked of the power of stem cells to cure cancer or overcome aging. But few hard facts backed up their rosy scenarios.

Foreign policy concerns increasingly crept into science and technology policy. The State Department complained that its staff lacked the knowledge of science and technology necessary to employ in international relations. Incidents of spying led Congress to debate whether to bar foreign scientists from American soil. The administration itself restricted exports of certain high-tech components to countries not on the most favored nations list. Health and Human Services grew concerned about America's lack of preparedness to ward off a bioterrorist attack and held several symposia to discuss remedial measures. Clinton himself mulled over the wisdom of destroying the last vials of smallpox. In defiance of the World Health Organization, Clinton refused to destroy the virus, claiming to keep it for research. More skeptical persons recognized its potential for terrorism and realized that some unadulterated specimens would be useful in creating an emergency vaccine supply.

Long-standing issues also resurfaced. An NRC report criticized the United States for lacking the computer power and model-building skills to construct an up-to-the-minute model of the global environment. Students continued to reject science and technology in schools in increasing number. AIDS groups called for a more aggressive campaign against the disease. Universities redoubled their earmark campaign. Nearly $1 billion went to congressional science and technology earmarks in 1999, an all-time high. John Silber, chancellor of Boston University, defended himself and his institution against those who decried pork barrel science. "People talk about sordid—they better read the Constitution," Silber said. "I've got a right to appeal to the Congress. Every citizen does. If that's dirty, then the whole country's dirty. The Constitution's dirty." And scientists continued to argue that earmarks not only took money from peer-reviewed and therefore more meritorious science, but also lowered the public's estimation of scientists to that of lobbyists.[54]

SUCCEEDING CLINTON . . . OR REAGAN?

The election of 2000 would be the first election since 1988 in which there would be no sitting president. Campaigns were in full swing soon after the first of the year. Both parties curried Silicon Valley's favor. It was a foregone conclusion that Gore would be the Democratic standard bearer. His long interest in science and technology made him a natural to the high-tech contingent. But some of Clinton's policies, such as his opposition to limiting securities lawsuits and to the entrance of large numbers of technically skilled immigrants to the United States, led techies to view him suspiciously. His continued and willful refusal to let the sequences uncovered in private laboratories be patented by individuals and then licensed to manufacturers struck them as a drag on venture capital and an assault on an inventor's individual property rights. No

Republican frontrunner galvanized the Valley's attention but the party's refrain of lowering restraints on taxes and trade as well as opposing federal dollars from directing technological innovation made any candidate likely to secure the nomination popular.

Clinton provided a boost to Gore supporters. Speaking at California Institute of Technology, Clinton revealed that his 2001 science budget recommendations called for a 7 percent increase in science spending. Almost all the new spending was to be channeled through grants for university-based research. He expected universities to produce "fundamental insights" as well as the next generation of scientists and technologists. His NSF proposal was more specific. He wanted a 17 percent hike, with the money going to "core" disciplines—condensed matter, materials science and the like—new multidisciplinary areas composed of physics, chemistry, mathematics, and engineering that he felt had been overlooked. This, of course, had been primarily Clinton's doing. His policies had first fostered technology and then health, medical, and other lifestyle sciences. Here he argued for "restoring balance to the federal R&D portfolio," claiming that the American scientific and technological enterprise had become warped and required redress.

Clinton argued that America stood at the onset of "an era of unparalleled promise—fueled by curiosity, powered by technology, driven by science." To reach that era, Clinton also proposed a large increase for Gore's favored information technology. Creation of a National Nanotechnology Initiative marked a departure from past practice. He hoped to investigate the dynamics of minuscule biological or mechanical devices, which he anticipated would become a new area of economic growth. Aware that Japan and Europe were exploring that terrain, Clinton thought it essential to build a research base in that possibly quite fruitful sector. His budget made a similar argument for biocomplexity, a philosophy that considers how simple life systems organize themselves into more complicated organic networks.

Nanotechnology and biocomplexity were Clinton's proposed scientific investments. With typical hubris, the White House science office proclaimed that nanotechnology "will lead to the next industrial revolution." Myles Brand, president of Indiana University and chair of the Association of American Universities, agreed. He claimed, "We are reaping the rewards of the money we have invested in university research." But Brand feared for the future. "Japan and Germany are spending a greater percentage of their GDP on basic research than is the United States, and South Korea is approaching our level," he warned. Only by greatly accelerating the spending pace could America guarantee preeminence.[55]

Such an argument ignored the fact that in gross dollars America spent much more than double those three nations combined. It also ignored the 1995 budget agreement between Clinton and Congress. There, deficit reduction had taken precedence. With that raison d'être gone, many Republicans in Congress wanted to pass the surplus predicated on that agreement on to their customers—taxpayers—by reducing taxes. Most Democrats took a different view, arguing that deficit reduction caused numerous items to be sacrificed. Now those domestic programs, including science, could be addressed.

This was first-class election year rhetoric. The gauntlet was cast down. The two parties defined themselves as clearly as possible. The only thing that remained for them to do was to spar for a few months and then to fashion a compromise. That compromise agreement would allow them to both declare victory, to show their constituents that their respective action produced measurable, positive difference and therefore merited continued fervent support. Well before that time, a certain amount of jockeying needed to occur. Defense put in its bid for a heightened appropriation as did Energy. Scientists complained about the necessity of undergoing rigorous scrutiny before receiving a federal science post and the prohibitions upon leaving government service. They also denied that procedures to prevent spying in scientific and technological matters were in most cases necessary. Universities maintained that they had undercharged government millions of dollars in overhead expenses.

The months passed and no compromise emerged. Both sides held to their guns and both pointed to their nobility of purpose. In the meantime, George W. Bush, the son of President George H. W.

Bush, captured the Republican nomination. As governor of Texas, Bush ran on a record of being a uniter, not a divider. He held strong conservative business principles, but called himself a "compassionate conservative," someone willing and able to assist those in need.

As had become the custom, *Science* interviewed both men separately and ran the answers to the same questions concurrently in a fall issue. Gore continued the agenda that Clinton had inaugurated in the way that Clinton had. High-tech jargon and forward-pointing rhetoric joined together to unveil new programs and new agendas. Bush was less focused and less directly responsive. His vague, brief statements proved more suitable as campaign slogans. Repeatedly he alluded to Republican principles, talked of the need to educate a workforce for a high-tech future, and stated that he favored free trade and lower taxes. While acknowledging he had no obvious scientific credentials—Bush held a master of business administration—he pointed to Texas under his stewardship as a technological marvel. More high-tech start-up companies were founded there than in any other state.

Gore also brought forth the standard list of scientists and technologists. The initial phases of the congressional science budget compromise provided further assistance to both campaigns. NSF received a 13.3 percent increase, a major victory for the Gore camp, but the funding that Congress refused to provide came from information technology, nanotechnology, and biocomplexity, a significant Gore defeat. It also tacked on a measure to bolster the research environment in those states that received the fewest federal science dollars. By setting aside a specific sum for those states, Congress hoped to balance out more nearly appropriations to different sections of the nation. Congress also passed a NASA budget far in excess of what the Clinton administration proposed, a clear victory for the defense-oriented Bush.

The rest of the budget remained unresolved until after the election. In late November Congress reconvened and by mid-December had passed the remainder of the 2001 science budget. Citing Bush's election win, some Republicans proposed freezing the budget until the inauguration. Various science advocacy groups marched on Washington to protest that scheme. In the end, science funding reached new heights. NIH was a major beneficiary, receiving 14.2 percent in new appropriations. Other branches received a bit less but the overall trend was the biggest increase for science funding in over two decades. Rather than rest on its laurels, *Science* wondered of the future. Its headline screamed: "Record Year for Science, but Can It Be Repeated?"[56]

THE DISLOYAL OPPOSITION

Congress also cared for constituents as it waited for Bush to take office. It authorized $1.7 billion in science earmarks, mostly to universities to build new science and technology facilities. Clinton also stayed active. He cemented his legacy by defining government science as multicultural when he approved within NIH a new National Center on Minority Health and Health Disparities. Supported by Senators Kennedy and Bill Frist (R-North Carolina), a physician, the center provided grants to investigators to study questions of difference in health and forgave student loans for researchers examining similar issues.

Partially because of his platitudes and vague statements of free-market principles, scientists viewed the Bush administration suspiciously. That they found Clinton and other Democrats more receptive to notions of using federal funds to support a wide variety of scientific activity accounted for some trepidation. Bush's brief candor during an unguarded campaign moment certainly worried them. When asked about government, Bush claimed that governments by their very nature were insatiable and would spend whatever was allocated to them. His idea was to reduce drastically through massive tax cuts the funds available to government. Only through economic starvation could he compel government to be efficient and lean and to concern itself with its legitimate responsibilities.

Within the Beltway, this sentiment proved disturbing. A rumor quickly circulated. The Bush administration had fired NSF director Rita Colwell. A nonpartisan appointment with a six-year term, Colwell was only halfway through her term but had publicly fought with congressional Republicans over the NSF director's power. Immediately the National Science Board, whose job it was to oversee the agency, contacted various politicos and warned them about "politicizing" the NSF and by extension, science. (Apparently, the irony of politicians contacting politicians to tell them not to be political was wasted on the actors in this drama). Scientific societies mobilized and collected signatures demanding that the new president "maintain the independence" of the NSF as an important safeguard for the "integrity of basic research."[57]

All those in a position to know denied that Colwell's job had ever been threatened or might be. Nonetheless, the wariness of the scientific community toward Bush made it conceivable. The first glimpses of his science budget furthered discomfort. Published as *A Blueprint for New Beginnings*, Bush proposed increasing the NIH budget by 14 percent but wanted to hold the line on science programs at NASA and the Departments of Energy and Interior. Special funding for nanotechnology and information technology would disappear. The NSF would get a small increase but that money would go to fund Bush's new science education initiatives. As a campaign promise, Bush pledged to ensure accountability in public schooling. He challenged the NSF to oversee that program. Bush's budget also refused to increase the size or duration of NSF grants, two leading agency objectives, or to fund construction of new facilities.

Critics noted a disturbing trend. Doctorates in the physical sciences granted in the past year had been the fewest since 1991. Biology PhDs had risen 30 percent. They postulated that a crisis was just around the corner; as physical scientists reached retirement age, there would not be enough recent graduates to replace them. Bush's budget exacerbated the situation and would further cripple physical science.

One-year survey results mean nothing, of course. Just a few years earlier, the various scientific societies worried about the lack of employment opportunities in the physical sciences. Bromley brought the NIH's success into sharper focus. "Obviously, every congressman . . . understands NIH," he sniffed. "Sooner or later, they know they're going to be stark naked on a gurney looking up at a doctor."[58]

Bush's unwillingness to reach out to scientists proved particularly unsettling to those with easy White House access. His appointment of E. Floyd Kvamme as a key technology advisor frustrated them. Kvamme was a venture capitalist, an expert in trade policy, and a strong proponent of Bush's educational goals and tax cuts. Bush summed up his attractiveness succinctly. As an industrialist with long experience in Silicon Valley, Kvamme was "a risk-taker. He understands risk and reward." Bush followed up Kvamme's selection by announcing easing restrictions on overseas sale of high-powered computing equipment, a longtime goal of Silicon Valley businessmen. Bush's act reflected his philosophy of governance. Government exists "not to create wealth," but creates an environment in which the "entrepreneurial spirit can continue to flourish."[59]

Bush further alienated himself from scientists generally by reneging on his campaign promise to decrease carbon emissions. Citing fuel shortages in California and other western states, he refused "to take action that could harm consumers." Donald Kennedy, then editor of *Science*, attributed Bush's change of position not to concern for consumers but rather because industry spokesmen had his ear. Kennedy bemoaned that Bush had not chosen to surround himself with scientists. In fact, Kennedy suggested that Bush's lack of balance in the science budget stemmed from his lack of science advice as did his position on global warming. It was time, Kennedy wrote, for Bush to appoint an "authoritative science voice" to tell "those who argue that global warming isn't to be taken seriously, 'Mr. President, on this one the science is clear.'"[60]

Four months into the Bush administration there had been no high-level science appointment. At an MIT gathering to celebrate the twenty-fifth anniversary of the OSTP, rumors circulated that Bush was considering closing the science office as a vestige of time past. The several former science advisors present took the news hard and sent Bromley and David to try to convince Bush to maintain the office. They also began to play the economic card. Marye Anne Fox, Bush's science advisor while governor of Texas, noted the close connection between science funding and the economy. Without an infusion of new federal science funds, she argued, "innovation and our competitive edge will shift overseas." Bromley made the economic case for generously funding federal science less lavishly. "No science, no surplus. It's that simple."[61]

Bush promised to reconsider the matter. Other rumors made the rounds. A report circulated that an EPA document demonstrating a linkage between dioxin—a constituent of Agent Orange—and cancer had been overturned at the chemical industry's behest. To be sure, an agency report linked dioxin to cancer. But an independent scientific advisory panel within the EPA had cast serious doubt on the science in the report, especially the assumptions used for the risk assessment model. As a consequence, it recommended wholesale revision before placing the report in the public record. Within a couple of weeks, the revisions had been made and the dioxin report released. Industry chemists criticized it for using unsettled science and contended that little was known about dioxin and humans.

Increasingly, scientific and technology organizations and groups looked beyond the White House for solace and influence. Congress emerged as a friendly face. Sherwood Boehlert (R-New York), the new chair of the House Science Committee, took the lead. Claiming to be "a proud, card-carrying moderate," he pledged to operate his committee "in a way that would make Einstein smile." Scientists responded to this boast by thanking him for not "playing dice with our universe," a reference to what Einstein had said God would not do. Boehlert pronounced himself disappointed by the Bush science budget and promised that next year's would be better.[62]

In the meantime, the Bush administration continued its policies of dismantling or reassessing the programs and initiatives of the Clinton administration, and of establishing its own precedents according to its principles. As president, Bush was philosophically more like Reagan than he was like his father. He viewed much of what government did with suspicion, as something best left to private enterprise. His presidential appointments reflected that intellectual kinship. He reached over his father's administration to select persons experienced in the Reagan administration. Secretary of Defense Donald Rumsfield and Vice President Dick Cheney emerged as leaders of this neo-Reaganite group. But unlike Reagan, Bush proved a much more stubborn sell. Reagan could be convinced of certain needs and obligations; Bush proved much less tractable, unwilling to admit possible mistakes or even that he changed his mind.

Bush wasted no time putting his plans into action. He gave high priority to the remnants of Reagan's Star Wars missile defense system and experimented with different ways to disable incoming missiles. He thought the basis upon which America had structured nuclear deterrence—Mutually Assured Destruction—was foolhardy, and he wanted to establish an umbrella to prevent enemy missiles from raining on the United States. This required him to announce plans to abrogate the 1972 Anti-Ballistic Missile Treaty, a cornerstone of MAD, and to seek technologies to shoot down missiles as they initially entered space. That specification necessitated sending interceptors over the air space of Russia and China almost immediately after a rocket launch. There could be no mistake, no second chance, with this sort of defense. Similarly, the administration quickly opposed the Kyoto Treaty, arguing that global warning was speculative, not a proven concept, and therefore it would be folly to risk American economic prospects. A Vice President Dick Cheney–chaired report on energy policy favored expanding resources over conservation and suggested opening the Alaska refuge to drilling as well as investing more heavily in nuclear power. Significant new funding for fusion and superconductivity also gained the Cheney panel's support.

Six months into his term, Bush selected a science advisor. John H. Marburger fit the profile of previous science advisors. He was a physicist and had experience in scientific administration. As important, he offered strong support for administration positions. He acknowledged that science was not going to resolve the stem cell question, that the economic implications of Kyoto were "kind of scary" and required more research, and that MAD as a strategic defense position was "bordering on insanity." Marburger pronounced himself pleased with the president's willingness to seek his advice as well as Bush's "acute grasp of the fundamentals" of science policy.[63]

Almost immediately Bush announced his first science policy decision. Despite moral and other reservations, he decided to authorize the use of federal money for stem cell research. But Bush included free-market principles within this decision. Federal support could go to only those self-sustaining lines—estimates ranged from twelve to sixty-four worldwide—for which the individual property rights had long been assigned. In this manner, Bush could claim to be part of no additional fetal destruction while honoring the right to hold biologically based intellectual property, an approach at odds with his predecessor.

Scientists generally expressed disappointment at Bush's stem cell initiative. Some maintained it was too limiting, likely to hold up important research. Others contended that it held science hostage to religion as Bush's compromise aimed to please the pro-life lobby. Ethicists urged public debate and asserted that the question when trying to balance the rights of all concerned was quite complicated. The NAS called for more lines to be made available to speed research.

9/11 AND THE WAR AT HOME

Each of these matters became markedly less significant on September 11, 2001. Hijackers flew three airliners into the World Trade Center and the Pentagon. A fourth crashed into the Pennsylvania countryside. Science and technology found immediate utility. Satellites photographed Ground Zero and a software package, GENIE, helped identify subtle patterns on the surface through otherwise visually impenetrable debris. Environmental hazards as well as suspicious sites could be detected from the air. Almost immediately, a new range of science and technology issues emerged. How was the nation to protect itself from terrorists willing to use weapons of mass destruction against civilian populations? Could there be ways to make America safe? A second assault quickly seemed to hit the nation. Anthrax began appearing in letters sent to several congressmen as well as the *National Enquirer*. Biological and chemical weapons suddenly seemed real and present.

The crisis transformed Bush's public persona. His inability to gain hold of the reins of government had made him unpopular. He struck many as awkward, cocky, uneducated. His slavish adherence to his fundamental principles had seemed to a significant portion of the electorate to injure America and to weaken his office. He apparently lacked the charisma of Reagan and the thoughtfulness of his father. But with the unthinkable about, he forcefully seized command and Americans rallied around their leader.

Science and engineering helped out. The NSF funded robots to search Ground Zero for survivors. DOE chemical and biological experts hunted the anthrax poisoner, albeit unsuccessfully. The NIH worked on a smallpox vaccine. Any number of terrorism task forces were created in the wake of 9/11. Almost all had to do with science or technology in some capacity. Why did the towers fall? What intelligence had not been intercepted and why? What scenarios might result from a massive chemical attack? How could one protect vulnerable structures, such as nuclear power plants? Congress, the NAS, and the Defense Department all called for reform. In a top secret meeting, the NAS, the Institute of Medicine, and the National Academy of Engineering invited leading scientists to Washington to recommend to the White House how science could combat terrorism. Among other things, the conclave listed possible terrorist targets inside the United States and called for new research areas in scientific antiterrorism. Much of the deliberations smacked of the scenario strategy that had long been an integral facet of American foreign and defense policy. Scientists and engineers would plot out what to do in case something happened.

Such an approach is bounded by the limitations of the individuals involved. Congress sought to add its own initiative by requiring registration and accountability of biological agents and hazards that could be used as weapons. Researchers chafed at the suggestion that dangerous bioagents needed closer scrutiny, arguing that studies would be slowed and benefits delayed. But investigators were surely cheered when Congress ignored Bush's budget and in the wake of the new war on terrorism provided a hefty 7.7 percent increase to the NSF. Surprisingly, Congress left the NSF budget free of virtually all earmarks; agency administrators gained great latitude to decide to which projects funding would go.

The Bush administration embraced science and technology as counterterrorism agents. It sought assistance in peering through walls and detecting bombs. It wanted new plans to protect power grids and computer networks. It needed devices to track bio- and chemical weapons. It sought the insight of social scientists to peer into the terrorists' minds. The OSTP was given the responsibility of providing the Office of Homeland Security technical support. That task included evaluating the various unsolicited suggestions from the public of how to wage war on terror. Even the next year's budget proposal (for 2003) placed aside an additional $3.4 billion for science and technology in service of antiterrorism.

Bush's antiterrorism science seemed also to produce a softening in other areas. Perhaps because Bush understood the need to engage the international community, he produced a plan to curtail greenhouse gas emissions in the United States. Scheduled to reduce emissions 18 percent by 2012, the Bush program relied on market forces —$4.6 billion in tax credits—to achieve the goals. If the nation failed to reach that level, the administration promised to enact draconian measures. Bush also completed numerous appointments of midrank science personnel. Associate directors of the OSTP, a science director of the DOE, and a beefed-up EPA science team signaled new utility for science within the administration. Yet it would be wrong to suggest that tension between the administration, scientists, and engineers had abated completely. Classified research rested at the heart of a continuing disagreement.

Marburger set the tone. He thought it prudent for Homeland Security to require academic scientists to withhold research in areas such as recipes for toxic chemicals or biosequences of potential pathogenic weapons. Duke and MIT claimed it went against their principles to undertake work that did not permit a full and free exchange of information. Other universities wondered if foreign nationals and others should be kept from certain research areas. Marburger's solution was simple. Each campus should erect its own rules. Homeland Security then could choose to place research only in universities with welcoming policies.

Congress had a different idea. It required all biological facilities (an estimated 190,000 places) to report if they experimented with or possessed any of thirty-six pathogens easily converted into biological weapons. That law was followed up by another that focused on agricultural chemicals that might be used to produce nerve gases, toxins, and the like. Only researchers with "a legitimate need" for these agents should have clearance to experiment with them. The American Society for Microbiology and several universities supported the principle of governmental regulation of scientific endeavor but disliked "the incomplete and ambiguous" regulations specified in the law. Individuals argued that placing restrictions on pathogens within the states was of little consequence because they could be brought into the country rather easily. Others maintained that restrictions would boomerang; antidote and vaccine production would be slowed by red tape. Still others worried about individual investigators who needed to satisfy tenure deadlines. Delays would hamper their ability to secure tenure.[64]

Focus on biological and chemical agents extended to those who worked with them. The State Department began to scrutinize much more extensively scientists or engineers hoping to enter the United States. These people were by definition potential menaces; they had the ability to engage in bio- or chemical terrorism. Visas were regularly denied and meetings and conferences canceled. This was true even when those conferences were about how to defend a civilian population against bioterrorism.

About a year after 9/11, the Bush administration issued guidelines to control sensitive information as well as access to that information. Assuring scientists that most information would remain open and that prepublication review by the government would only be restricted to a relatively few cases, Marburger also expressed concern "about inadvertently giving aid to the enemy." To enforce these regulations, he announced creation of a new federal board, the Interagency Panel on Advanced Science and Security, to systematize what would be censored and what might be released.[65]

Security had become an ingrained part of the fabric of American science. To coordinate the various security efforts, Congress formed the Department of Homeland Security. This cabinet department combined the efforts of twenty-two existing government agencies and established new security programs. Science figured prominently. An undersecretary for science and technology, operating under a twenty-member advisory board, oversaw those areas. Among the new initiatives was a Homeland Security Advanced Research Projects Agency, which awarded contracts for detection devices and border security techniques; a Homeland Security Institute, a think tank to draw out possible scenarios and means to defend against them; and an advisory role in the NIH's new $1.5 billion bioterror research program as well as a similar USDA program. Congress left these life science programs within established life science divisions, arguing that expertise already existed there and that those programs were synergistic with others in the NIH and the USDA.

Elevating Homeland Security to cabinet rank empowered the department and further frightened researchers. The unquestioned and unabashed deference to the White House that had emerged in the wake of 9/11 had all but evaporated as scientists and the Bush administration renewed antagonisms. A panel of prominent scientists drew attention to what they claimed had become the disproportionate funding of the biologically based NIH and the relative neglect of the physical sciences. A call for balance had begun during the Clinton administration but those hoping to convince Bush saw the matter more starkly after 9/11. Concern over biological and chemical agents had led to an even heavier NIH investment. By their accounting the physical sciences rested some $16 billion per year behind their biological counterparts.

The group took their findings to Marburger, who proved unsympathetic. He challenged then to document the injury to the physical sciences and maintained that neither he nor the president held any stock in arbitrary formulations. Balance was one such construct. So too was any reference to something like "doubling" the budget or by setting a specific time period to achieve that end. If the group had an important rationale to argue for increasing physical science moneys, he and the president would listen, but vagaries and stock complaints would get no hearing in the Oval Office.

In truth, physical sciences always got and would always get the lion's share of the federal science and technology budget. Much of the defense and military science spending was in the physical sciences. What the panel and others were complaining about was discretionary or "basic" science spending—nonmission directed science. In that sense, their analysis may also have been inaccurate. While NIH spending had indeed escalated, it did so in particular targeted areas: AIDS, biohazards, and other applied or earmarked subjects. The vast majority of funding was not curiosity-driven but rather to settle an immediate problem. Also of significance was the manner in which Marburger responded to the plea. Here he was reflecting how Bush reasoned. In that sense, Bush was not at all like Reagan, who could be swayed relatively easily. Bush was far more dogmatic and doctrinaire, perhaps like a hardheaded businessman who justified everything and anything by the immediate bottom line. Results needed to be swift and problems pronounced if Bush was going to invoke the authority of the government.

Despite an economic slump exacerbated by 9/11, an invasion of Afghanistan, and an impending war in Iraq as well as a budget deficit likely to top $300 billion, the Bush administration accelerated science and technology spending in its proposed 2004 budget by almost 10 percent. Relatively little of the new money went to discretionary spending. The vast majority went to military- and homeland security-related science efforts, although the NSF also did well. But the environment, NASA, and climatology all took hits. Areas Bush deemed less important, perhaps even frivolous, suffered stagnation or even decreases.

American universities reflected the new trend. While some years ago, much of university research revolved around commercial applications, now it centered on security. How to detect biological weapons supplanted how to identify a disease. Defense spending on campuses had risen some 27 percent during the past three years, while NIH spending rose 31 percent. To be sure, 9/11 had a predominant impact. Perhaps the most dramatic example of a post 9/11 scientific endeavor was the creation of regional biodefense research centers. Located at eight universities scattered across America, each focused on the development of therapies, vaccines, and diagnostics to combat potential biological terrorism. But so too did Bush priorities. Innovation and application were private endeavors, except in terms of the national defense. Government funds science and pre-emergent technologies. Private enterprise operationalizes all else.

Precious few new scientific or technological initiatives outside of homeland security were sponsored in the first years of the new century. The Next Generation Lighting Initiative was a major exception. A consortium of DOE labs, academia, and private enterprise, the initiative worked to use light-emitting diode and organic thin films technologies to produce solid-state lighting. Its proponents argued that this lighting would capture the global market—Japan and Korea were working on the process—reducing energy consumption by more than 10 percent.

SCIENCE OR POLITICS?

Paralleling the rise of post 9/11 science was the demise of independent advisory groups within the government. The Federal Advisory Committee Act required governmental advisory bodies to "be fairly balanced in terms of the points of view represented and . . . not be inappropriately influenced by the appointing authority or by any special interest."[66] But according to a *Washington Post* news story, the Bush administration failed to follow both the spirit and the letter of the law. It disbanded the National Human Research Protections Advisory Committee and the Advisory Committee on Genetic Testing because they offered views at odds with certain Bush religious constituencies. It even replaced fifteen of the eighteen members of the advisory committee for the Center for Environmental Health, many with representatives from the petroleum or chemical industries. *Science* saw Bush's advisory panel selection as a charade. Rather than note and attempt to grapple with "scientific ambiguity," advisory committees were to support "the administration's antiregulatory views." And in situations of "deeply held conflicts in values, we can expect only silence." The goal of the administration, the authors claimed, was simply "regulatory paralysis."[67]

The issue of scientific advice was a thorny one. If the matter was not in dispute, then there was no problem and no need for advice. But no one was suggesting a scientific plebiscite; no one wanted the majority to rule, no matter what they said. So the terms of the debate rested on "best science." That was the rub. Best science rarely was the same to the various stakeholders. When things were not crystal clear to all partisans, then philosophical underpinnings—ideologies—ran rampant. At least since Nixon and perhaps since the 1964 Johnson-Goldwater election, there had been an open acknowledgment that approaches to governance, to society, to life mattered in what one expected from science. It colored what was scientific proof. It led to choices that caused disputes. In short, it made science unscientific, unable to resolve any number of questions.

To be fair, Bush advisory panels did agree on some things that *Science* might have acknowledged. An OSTP panel suggested small-scale premarket testing by the EPA and the FDA of genetically altered crops to check on their probable safety before undergoing field tests so large that contamination with regular crops was all but unavoidable. Yet that decision was swallowed up within a larger debate about genetically engineered food. At precisely the same time that protestors assembled to denounce genetically modified foods, USDA secretary Ann Veneman championed the products as necessary to feed the global population. Bush himself argued that the boycott of genetic foods by European nations stemmed from "unfounded scientific fears." Marburger went further. He characterized resistance to this new food technology as willing to sustain a perpetual "vicious cycle of ignorance and poverty."[68]

Scientists in their conclaves overwhelmingly voiced their opinion that regulatory and advisory matters were being handled politically rather than scientifically. Even the National Academies chipped in. They created an ad hoc panel to analyze the Bush administration's "capacity to select highly qualified individuals for the top science and technology-related advisory committees." A direct swipe at Marburger and others, the ad hoc panel would "assess the current recruiting environment" and review policies aimed at "ensuring that the panels are balanced and independent." It threatened to recommend to Congress "changes in the recruiting process" if it found that highly qualified scientists were not being chosen.[69]

Democrats in Congress joined the crusade. Speaking for the minority party on the House Governmental Reform Committee,

Henry Waxman (D-California) issued a report outlining some twenty alleged Bush administration abuses of the scientific process "far beyond the typical shifts of policy that occur" when a new party captures the White House. "Political interference with science" included "misleading statements by the president, inaccurate responses to Congress, altered web sites, suppressed agency reports, erroneous international communications, and the gagging of scientists." Only one factor binds all these abuses together: "the beneficiaries of the scientific distortions are important supporters of the president, including social conservatives and powerful industry groups." Bush's practice has resulted in "unprecedented criticism from the scientific community." The current administration, Waxman decided, "has repeatedly suppressed, distorted, or obstructed science to suit political and ideological goals."[70]

At least one policy analyst saw the matter differently. Roger Pielke Jr., University of Colorado science policy analyst, did not understand the terms of the debate. Politicians engaged in politics; they adhered to philosophical positions about the nature of governance. Why should we expect "political advocates . . . to resist seeking to influence political outcomes by manipulating science?" he wondered. If we did that, he reasoned that it was "also fair to expect the scientific community to resist using science to seek desired political outcomes." "From a political perspective," Pielke argued, "science is used to reduce choice among decision-makers . . . to a single preferred outcome." "Too often," he concluded, "members of the scientific community conflate policy and politics"; they "set the stage for the politicization of science by scientists and advocates alike."[71]

Others thought that the matter revolved around the lack of balance in science advice between the White House and Congress. The situation stemmed from an event in 1995. The new budget-slashing Congress killed its Office of Technology Assessment, leaving the branch to depend on the soon overwhelmed NAS, which offered only one alternative, and the politically charged OSTP. It neither received policies, options to sift through, nor the deep, rich understanding of a scientific controversy that a division committed to that project could provide. But even that became quickly politicized. A bill to recreate the

Office of Technology Assessment was introduced into Congress by Rush Holt (D-New Jersey) as a critique of the Bush administration. Not surprisingly, the Republican leadership opposed it.

Ironically, as the scientific establishment and the administration battled, a July 2003 Cato Institute study argued that the Bush administration had issued a record high number of pages of federal regulations during the past year. At first glance, more pages of regulation may suggest a closer monitoring of the environment, food, medicine, and the like. But it could also indicate wordier, more convoluted regulations, containing caveats and other exceptions to undercut the spirit of the rules. In that sense, additional pages could equal less vigilance and more exceptionalism.

Armed with that report, the Bush administration proposed a new policy to oversee regulatory rules. Agencies seeking to implement new rules must convene panels of independent experts to review "the quality of science" in suggested regulations. In this manner, the administration contended, lawsuits could be avoided and a "more consistent regulatory environment, which is good for consumers and business," would be established. Kurt Gottfried, leader of the Union of Concerned Scientists, hoped that the new procedure "would prevent the kind of abuses that the administration has engaged in pretty systematically," but he claimed to be "pretty skeptical about what the intention is here." The Cato Institute's Jerry Taylor disagreed with the whole project. Acknowledging that many persons, including scientists, believe "that if we just get the science right then we can more easily adjudicate" regulatory matters, Taylor argued instead that such a perspective was "just naïve." It failed to recognize that these questions and decisions were often political in nature.[72]

The scientific critique of the Bush administration dissipated during later fall and winter, 2003–2004. Its mercury emission reduction plans offended environmentalists but with few public repercussions. A DOE panel congratulated the agency on its scientific initiatives but complained that these endeavors remained hidden. Enhanced visibility would result in increased support. The agency complied by beginning two new high-visibility projects, the International Thermonuclear Experimental Reactor and the UltraScale Computing Facility. Supercomputing and controlled nuclear fusion had long been DOE staples. They now were to become focal points. Congress also made its presence felt. It exempted Defense from following certain environmental rules, including the Endangered Species Act.

Bush even attempted to set a grand new scientific venture to serve as an administration focus. America would go to Mars. Bush envisioned a permanent moon camp to act as a fuel depot and manufacturing site for the trip to the red planet. This initiative, he continued, would yield several benefits. Defense and NASA would work together closely—a new rocket booster might be useful also for intercontinental ballistic missiles—therefore reducing duplication and increasing efficiency. A Mars landing could lead to new technologies, including new space materials for clothing and to help combat zero gravity, and uncover potential energy sources. To help pay for the hefty $500 billion estimated price tag, Bush proposed reallocation of resources, including phasing out the international space station and the space shuttle. (See fig. 11.)

Bush's proposal played to a mixed audience. Some touted the vision while others disputed various aspects. Its cost while considerable social issues remained on earth was one objection, as was the risk to human life. Others saw manned space travel as needlessly expensive. Unmanned flights were much cheaper. But Bush's proposition captured the scientific community's attention only briefly. Gore helped rekindle the anti-Bush scientific animus. He lambasted Bush for his failure to listen to what Gore claimed was a nearly unanimous opinion among scientists on the threat of greenhouse emissions and global warming. For the latter part of 2002 and all of 2003, Bush had ignored virtual scientific consensus, Gore noted. Instead, his administration had participated with "wealthy right-wing ideologues" and "the most cynical and irresponsible companies in the oil, coal and mining industries" to "finance pseudo-scientific front groups." These groups "issue one misleading 'report' after another, pretending that there is significant disagreement in the legitimate scientific community in areas where there is actually a broad-based consensus." Even worse, regulatory and advisory committees had been

Fig. 11

In 2003 NASA launched a mission to Mars that carried two exploratory rovers. These robotic vehicles rolled slowly across the planet's surface, photographing the Mars landscape and analyzing its geology. The tough little machines outlasted their officially projected lifetime and drew a record number of visitors to NASA's Web site, as the rovers collected intriguing evidence about the possible past water activity on Mars.

http://en.wikipedia.org/wiki/Image:NASA_Mars_Rover.jpg

polluted. The administration thought nothing of "appointing the principal lobbyists and lawyers for the biggest polluters to be in charge of administering the laws that their clients are charged with violating." By turning to these campaign contributors the Bush administration was now "wholly owned by the coal, oil, utility, and mining industries." Bush himself was a "moral coward" for choosing campaign contributors over what clearly was the public interest.[73]

Others quickly rejoined the fray. Sixty scientists, including twenty Nobel Prize winners and nineteen National Medal of Science recipients, issued a protocol accusing the administration of politicizing science. "When scientific knowledge" conflicts "with its political goals," the group wrote, "the administration has often manipulated the process through which science enters into its decisions." In its thirty-seven pages, the protocol spelled out the instances in which it found Bush violating scientific assessment for partisan gain. It called for congressional action to ensure adequate scientific input in all regulatory and advisory matters.[74]

Marburger defended the administration. He said the document was "deeply flawed," based on "troubling misconceptions," and read like "a conspiracy theory report." Labeling the contents "disappointing," Marburger blamed himself for failing to get the administration's message across to his scientific brothers and sisters. He maintained that there was "no evidence to support this sort of sweeping condemnation of administration policymaking." Bromley joined Marburger in defense of Bush. He found the protocol a "very clearly politically motivated statement [containing] broad sweeping generalizations for which there is very little detailed backup." At least one scientist found himself unmoved by Marburger's denial. "I actually feel sorry for Marburger," noted Howard Gardner, a Harvard cognitive psychologist, "because I think he probably is enough of a scientist to realize that he basically has become a prostitute."[75]

In early April 2004 Marburger published a formal rebuttal. Claiming that his detractors were guilty of "errors, distortions, and misunderstandings," his seventeen-page defense included point-by-point refutations of the previous charges. But Marburger's main arguments in his pamphlet were of a quite different character. He argued that scientists should not expect to decide policy battles, that to think political operatives would accept science as the last word was not realistic. "Even when the science is clear—and often it is not—it is but one input into the policy process." With that caveat, he concluded that "in this administration, science strongly informs policy"; it does not dictate it.[76]

Marburger also promised to attend the annual meeting of the AAAS later that month to further discuss science–White House antagonisms. Bush's proposed five-year budget outline assumed center stage there. Al Teich, AAAS director of science and policy programs, led the critique. He claimed that the budget would reduce financing in twenty-one of twenty-four federal agencies involved in extensive scientific activity. Only domestic security, the military, and space exploration would be spared. "Particularly during a presidential election year, it is essential" that voters understand the thrust of these policies, Teich maintained. Marburger argued that spending for science under Bush was at historic highs and that beyond the budget for 2005, which called for no increase, the rest of the projections were speculative, dependent upon a rebounding economy.[77]

NASA may well have not had the privileged position in Bush's plans that its detractors believed. Rather than lead the fight for the Mars project through Congress, Bush left NASA leadership to make its own way. Not surprisingly, Congress was unmoved by the Mars gambit and scoffed at the initial $12 billion price tag. In the wake of the agency's failure to galvanize Congress, a White House panel—the president's commission on the Moon, Mars and Beyond—recommended that NASA remake itself if it hoped to turn Bush's space exploration vision into reality. NASA should emulate military models and the private sector and become a leaner, more focused organization. Its centers should each adopt dual use as the mission and compete with one another to find which means to an end achieved greatest efficiency. The committee also argued for a special division of space exploration within NASA so as to ensure that the Mars landing remained the primary objective of at least one agency branch.

Congress remained recalcitrant and turned to other science questions. There it found a chorus of criticisms of US science efforts. An NSF study reported that in some areas of science and innovation the nation had begun to lose global dominance. Its authors noted that articles appearing in *Physical Review*, a leading physics journal, written by American authors dropped in half and that the number of patents issued to American citizens was likewise slipping. Others looked to newer technologies, such as stem cells, hydrogen cars, and nanotechnology, and they saw insufficient funding to guarantee market domination in the future.

This and the preceding criticisms of the Bush administration had arisen during the first months of the 2004 presidential election. With his nomination as the Democratic candidate all but assured in late June, Massachusetts senator John Kerry took on Bush's science efforts directly. Armed with a letter from forty-eight Nobel laureates stating that Bush was "undermining the foundation of America's future," Kerry claimed that "America deserves a president who believes in science."[78] Bush has failed with AIDS, Alzheimer's, cancer, and Parkinson's, Kerry argued, because he allowed ideology to triumph over science; his ideological supporters rejected stem cells. If he became president, his first two acts would be, Kerry maintained, reversing the gag rule that required scientists within the administration to cleave to Bush's ideological perspective and opening up stem cell research.

The Union of Concerned Scientists leveled an additional charge later that week. It maintained that as part of the vetting process for federal panels, prospective scientific appointees were asked two inappropriate questions: where did they stand on the issue that they were to adjudicate? and had they voted for Bush in the previous election? The Bush administration admitted that that had happened in some instances but it maintained that the answers did not affect any appointments and that the practice was quickly stopped.

Even Congress got involved. Vern Ehlers (R-Michigan), a physicist by training, argued that who did you vote for was "an appropriate question to ask." Scientific panels did not function in a vacuum. "Scientists must study the policy process and willingly participate." Panelists "must be in touch, even in tune, with political realities around them." There must "be no role whatsoever for ideological or partisan litmus test," countered Waxman. "For political appointees, the president should expect that his nominee supports his policies. But for advisory committees, they ought not to ask one's views on abortion, or how they voted."[79]

The UCS took the campaign to the next level by having scientists sign a petition protesting Bush's treatment of scientific advisors. Approximately 4,800 had signed by mid-August. Another group, Scientists and Engineers for Change, planned to give a series of public lectures in ten closely contested states. Unlike the UCS group, SEC focused on science funding and prominence. "Science counts, and it has not counted sufficiently in this administration," claimed Vinton G. Cerf, an Internet pioneer. We hope to get the electorate to understand "the importance [that science and technology] has in our society." We are as a nation "at risk of losing the edge" and basic science budgets must mushroom to maintain the world leadership.[80]

While scientific groups opposed Bush and generally favored Kerry, technologists found little in either candidate to cheer them. Bush's first term in office provided few of the incentives that he had promised high-tech entrepreneurs—tax cuts for technology-related activity, waiving of antitrust legislation, and clarified intellectual property rights. Kerry was viewed no more positively. His long career on the Senate commerce committee resulted in almost no high-tech initiatives. There were very few pertinent remnants from his Senate work.

Science again undertook its quadrennial survey of the views of presidential candidates. When asked about his science and technology priorities, Bush favored increasing Internet access, research toward a hydrogen rather than a petroleum economy, and using science and technology to combat terrorism. Kerry chose balancing the scientific portfolio, changing stem cell research policy, and restoring luster to the office of the president's science advisor. Both favored Mars exploration but Kerry would send machines, not men. Bush shifted global warming discussion from controlling hydrocarbon emissions to how hydrogen would remove the question. *Science* acknowledged that on most other matters they "look like Tweedledum and Tweedledee," but warned readers that real differences existed in "core scientific issues: climate change, space, stem cells, and the Endangered Species Act."[81]

Science's use of the concept of core issues was miles different than what had constituted core scientific issues decades earlier. It was testimony to just how ingrained science had become in American governance and just how closely identified those who pursued science had become with the consequence of that pursuit. The points of contention were a handful out of literally millions of government-social interactions. Granted, some of the areas of disagreement were wedge issues, deal breakers, but not to the vast majority of the tens of thousands of scientists laboring with federal dollars on projects in American laboratories. Control, respect, authority, power more nearly encapsulated the dispute. It was more about winning than anything else. Climate change and the like were really about who should have the power to govern.

Bush's razor thin margin of victory concerned scientists, although not as much as one might have thought. Despite their heated, vehement opposition to Bush and their refrain that the administration required political litmus tests, precious few scientists and engineers verbalized any fear of politically inspired punishment from an administration they had claimed did little more than play politics. Some worried of flat science funding, while others wondered if after the election the administration "reaches out and engages [the scientific community] or goes in its own direction." "I think," noted Michael Lubell, American Physical Society lobbyist, "the scientific community is now perceived by this White House as the enemy, and that will make it harder to open doors." Boehlert urged rapprochement. "Shame on both sides," he chided. The rhetoric got a little bit excessive. The administration should "demonstrate a greater degree of interest in the opinions of the scientific community," and scientists should learn that "what they have been saying [about Bush] hasn't helped the profession."[82]

Marburger offered a different critique of the election. He argued that science lost and would have lost even if Kerry had won. "Science needs patrons," he reminded his scientific colleagues, "and our patron is society." One must remember to look at the big picture—the vast money afforded science, the vast governmental power accorded science—rather than any particular aspect. "If we are not careful," he noted warily, "the scientific community can become estranged from the rest of society and what it cares about." Establishing science as an election issue not only demystifies science but also it loses sight of the big picture, a picture where science receives huge funding, accolades, and great measures of autonomy. "I do not think it was good for science" to have become immersed in partisan politics.[83]

Marburger seemed prophetic when the lame duck Congress passed its budget for 2005. Citing the high cost of the war in Iraq, the war on terrorism, and the huge deficit, it reduced Bush's science funding requests. While the entire science budget increased a slight bit from the previous year, the bulk of the new and reallocated moneys went to defense- and homeland security-related science and technology. A small amount went to NASA for its Mars venture. For the first time in years, the NSF received fewer absolute dollars than the previous year, a nearly 2 percent drop. Prospects looked even bleaker for subsequent years.

Alan I. Leshner, chief executive officer of the AAAS, explained in *Science* what he sensed was the nation's new mood. An alarming trend to downplay science had emerged. Without any remediation, the situation would cascade. A desperate Leshner proposed a radical solution. Scientists must "build stronger partnerships with [science's] beneficiaries and patrons in the public." He especially urged "alliances with leaders in industry." Partnerships with those "who will use our products to advance the public welfare strengthens" our cause. Leshner admitted that it was not his idea to treat science as a commodity but rather the suggestion of "scientifically sympathetic members of Congress," who "advise us again and again" that "messages from constituents" provide the greatest "political leverage." He noted that "reaching out to the public is not a strong tradition for the science community," perhaps because it arrogantly supposed "that nonscientists cannot understand our work." But, he continued, "we really need grassroots support. Alliances with leaders in local industry have a special kind of leverage, and science/industry partnerships can convince government representatives of the need to support science and its use for the benefit of society at large." Rather than "lamenting our fate," Leshner asserted that scientists "can mobilize our natural allies—the people we serve—to convince our policymakers not to make the same mistakes" of not adequately funding science in the future.[84]

Leshner's sentiments and proposal were extraordinary. He urged his colleagues to politic for science rather than party, issue, or candidate. He embraced industry. He embraced science as product. He embraced the public. He embraced the political process. Few of these stakeholders had been acknowledged so prominently by a representative of the scientific establishment in decades. It brought to the fore several critical issues: what was science, what was the relationship between the activity and those who pursued it, what was the nature of government, what was the responsibility of the practitioners of science, and many more questions. Prospects of funding did not immediately improve. The president's 2006 budget proposal included a 1 percent reduction for science, a figure that ought to have been much higher but, according to Marburger, "the president really cares about science."[85]

Leshner's words seemed to have had little impact on his fellow scientists, who simply offered the refrains that they had for more than a decade. Bush's "inadequate investments in research" will "erode the research and innovative capacity of our nation," complained Nils Hasselmo, president of the Association of American Universities. The Federation of American Societies for Experimental

Biology maintained that the budget will "discourage our most tal-
ented young people from pursuing careers in biomedical research."
Some in Congress and those in the administration disagreed. They
pointed to the massive yearly funding for science from the federal
government and wondered what an appropriate return for that
investment ought to be. In particular, they focused on the NIH,
which had had its budget more than doubled in the past five years
to tackle AIDS, cancer, Parkinson's, and Alzheimer's. Frustration
over endless promises and budgets, coupled with a Chicken Little
syndrome that always proclaimed that the sky was falling, produced
no measure of cynicism. "We have planted" for years in the garden
of biological science, argued Michael Leavitt, Health and Human
Sciences secretary. "It's time for us to harvest the fruit."[86]

Marburger stepped up the debate when at the 2005 AAAS
meeting he challenged the organization's social scientists to apply
their techniques on the scientific community generally. Noting that
economists had used behavioral models to study things like how
retirement patterns would influence social security, he called for "the
social science of science policy . . . to grow up" and tackle a series of
questions. These would include the scientific community's "voracious appetite" for federal money and the "huge fluctuations" in
state support of universities. Marburger wanted data, facts, truth.
Rather than simply count R&D financing—something "unbelievably
oversimplified"—he wanted an analysis that incorporated a whole
slew of things: public sector investment, project output, market
share, comparisons to other industries, and the like. What passed
now as social science policy "has an amateur-hour flavor to it." He
implored social scientists to get on board, maintaining that working
on science policy should not require "a lot of money . . . funding is
not the rate-limiting factor in this equation." At least on this final
point, Marburger found opposition. Significant federal funding was
essential to the quest, argued Connie Citro, director of the National
Academies' Committee on National Statistics. "That's what drives
academics in any field."[87]

CODA

Citro's comments and much of the material in chapter 4 show just how different expectations for scientists and technologists are at the end of the twentieth century and the beginning of the twenty-first. Scientists and engineers look to the federal government almost exclusively for their funding and therefore their livelihood. Universities count on scientists and technologists to generate support for their individual operations and university facilities generally. In a time of rising tuition and decreasing state and local government support for higher education, federal dollars appear essential. In a highly competitive world, scientific and technological prominence may well translate into commercial opportunities. Certainly that argument has worked on the national level. Only by funding science and technological endeavors generously can America compete in the new global marketplace.

There is a sense of desperation, exasperation, and entitlement in this analysis that ought not escape notice. Scientists and engineers demand, threaten, cajole, plead, and lobby. Their agents—universities, science and technical societies, and Congress (all politics is local, after all, and science and technology means a steady income stream)—are equally insistent and unrelenting. Scientists, technologists, and their partisans seem voracious, their appetite for federal funding implacable. To be sure, the later twentieth-century politics of partisanship increased the intensity of feeling. By painting opponents as unacceptable options, as pernicious possibilities, political groups cemented allegiance among their affiliates. Paradoxically, as identification with person and interest grew, influence became attenuated. No alternative existed. Persons could not play one party off against another and since they were so locked into the fold, they could not wield power by threatening to abandon their group. And the party deemed pernicious by the group recognized no advantage in attempting to reach out and placate people who avidly labored against it.

The sense of the insatiability of science and technology for federal funding comes from another avenue. It seems so omnivorous because we treat it as if it were a monolith. But science and technology are in actuality a series of competing interests, which appear under one umbrella. In the case of scientists and technologists, the realization that their causes are enhanced by lobbying as a single unit has produced the perception that science and technology are never fully satisfied, that no matter how much funding they receive they still feel shorted. In this framework, for example, funding authorized for physical sciences competes with money given to medicine, engineering, and social sciences and each is a yardstick against the others. That is an effective strategy for presenting a budgetary case but it runs the risk of disillusioning the public, the final source of support. A fine line exists between insufficient funding and the appearance of gluttony. A science and technology that presents itself as a monolith runs the risk of crossing that line.

That sophisticated approach to capturing federal moneys is a far cry from the approach of a half century earlier. There, science and technology in the public arena were indeed monolithic; they were the physical sciences and their applications. Few groups recognized the inherent advantages in identifying with them. During the 1950s the various ways that science and technology could aid America in winning the cold war became apparent. Biological and other sciences and especially universities began to see the possibilities of federal dollars. *Sputnik* greatly accelerated that awakening. It gave proponents a marker to rally around and a tangible consequence of what could happen if they did not.

Johnson's Great Society and War on Poverty brought the social sciences prominently into the mix, much to the initial chagrin of the physical scientists. Johnson's election also solidified the connection between the scientific and technical community and politics. Never again would this community stand completely idle during a federal election. Nixon had the hubris to threaten the community by removing its visible presence from the White House, a maneuver in many ways more menacing than withholding federal dollars. Ford welcomed science back into the Oval Office, and Carter tried to make scientists and engineers cleave to radically different principles and constructs than his predecessors. His efforts to direct them led to almost universal disdain.

Reagan institutionalized the Big/Little Science dichotomy, less because of the nature of inquiry than as a consequence of his free-market principles. After the fall of the Soviet Union, Bush attempted to be all things to all scientists and engineers, even establishing a formal federal presence for technology initiatives. Clinton highlighted federally selected and financed technology as an engine of American economic prosperity, which caused scientists and the new Republican Congress to balk. He placated both groups by dropping technology and fostering science during his second administration. George W. Bush continued Clinton's differential funding of health-related research but after 9/11 homeland security and defense dwarfed all other scientific and technological concerns.

Such a condensed overview demonstrates only the barest of cleavages that science and technology policies have sutured together. Since Kennedy's decision to select only people, including scientists and engineers, who passionately believed in his agenda, partisan politics has been part of federal science policy. In the 1980s partisanship among scientists and technologists took on a new cast.

Political partisanship gave way to professional partisanship. Since that time, scientists and technologists have treated funding science and technology as the end, not a means to an end. Rather than produce something tangible or otherwise, the principal goal became to maximize federal largesse. Scientists learned to confuse what they did with who they were; science was conflated with scientists. A much more personalized agenda and enterprise replaced the broader pursuit of science of an earlier time.

As it has been applied, American science and technology policy enabled government to incorporate new parameters whenever they emerged. This remarkably flexible rubric has accommodated a lack of commonality, a lack of core values, a lack of consistent approach, and even a lack of discrete, identifiable, continuing factors. Initiatives to superintend new areas of the American economy are integrated. Disparate plans to protect America, its people, water, food, drugs, and environment coexist comfortably, if not always peacefully. Indeed, plasticity may well be the most important constant in a science and technology policy that purposefully lacks restraints and absolutes.

This plasticity is the direct consequence of a pluralistic world in which democratic elections are the primary policy-manufacturing events. An entrenched technocratic elite exists—actually several do—but it does not control the reins of governance. In fact, it is directly dependent on government, not the other way around. To be sure, government needs scientists, technologists, and policy makers. But in the here and now they cannot thrive without government. They have become creatures of government, attempting to manipulate it to their benefit but without any recourse when things do not go their way. The irony is that scientists and technologists think they are in control, that they have the answers.

Government is tightly bound to science and technology but not to particular scientists, technologists, or policy makers. The huge battles and vituperative comments of the past two decades during elections followed by the sudden reduction in tensions and quick resumption of the status quo when the contests were concluded speaks volumes about the nature of the relationship between the government and the professions. Scientists, technologists, and policy makers have remade themselves as federal supplicants no matter how much they espouse their autonomy. Whenever they pursue any course of political action, the future is now. Decisions made in the present determine their ability to conduct and pursue their objectives in the foreseeable future.

ENDNOTES

CHAPTER 1

1. Vannevar Bush, *Science: The Endless Frontier: A Report to the President on a Program for Postwar Scientific Research, July 1945* (Washington, DC: GPO, 1960). Republished to celebrate the tenth anniversary of the National Science Foundation.

2. Ibid.

3. US Constitution, Preamble.

4. Dwight D. Eisenhower, "What Kind of Government Ahead?" *Vital Speeches of the Day* 26 (November 15, 1949): 66–68; Harry S. Truman, "Address on Foreign Policy at the George Washington National Masonic Memorial, February 22, 1950," *Public Papers of the Presidents—Harry S. Truman, Truman Presidential Museum and Library,* http://www.trumanlibrary.org/publicpapers/index.php?pid=662&st=&st1. The culmination of all these activities was a speech delivered in Boston on August 25, 1950, the Boston Naval Shipyard's sesquicentennial, by Mr. Francis Matthews. The arch-Catholic secretary of the US Navy, Matthews called upon the United States to launch an

attack upon Soviet Russia to make the American people "the first aggressors for peace." See *New York Times*, August 26, 1950, p. 16.

5. *An Act to Promote the Progress of Science; to Advance the National Health, Prosperity, and Welfare; to Secure the National Defense; and for Other Purposes*, Public Law 507, *Stat.* 64 (1950):149; and M. H. Trytten, "The New Science Foundation," *Scientific American* 183 (July 1950): 11–14.

6. The story of Princeton's move toward federally supported research is explored in Amy Sue Bix, "'Backing into Sponsored Research': Physics and Engineering at Princeton University, 1945–1970," *History of Higher Education Annual* 13 (1993): 9–52.

7. Dwight D. Eisenhower, "Address before the General Assembly of the United Nations on Peaceful Uses of Atomic Energy, New York City, December 8, 1953," *American Presidency Project, Document Archive* (hereafter cited as *APPDA*), http://www.presidency.ucsb.edu/index_docs.php.

8. USDA administrator of the Agricultural Research Service Byron T. Shaw's Memorandum to Research Division Directors is quoted in Gladys L. Baker et al., *Century of Service: The First Hundred Years of the United States Department of Agriculture* (Washington, DC: GPO, 1963), pp. 391–92.

9. Garrett Moritz, "From *Sputnik* to NDEA: The Changing Role of Science during the Cold War," http://www.gtexts.com/college/papers/j3.html; *Hearings before the Committee on Labor and Public Welfare, United States Senate: Science and Education for National Defense* (Washington, DC: GPO, 1958); *Hearings before a Subcommittee of the Committee on Education and Labor on H.R. 10381, H.R. 10278 (and Similar Bills) Relating to Educational Programs (Part 3)* (Washington, DC: GPO, 1958). Also see Micah Rueber, "Reds Launch First Moon," delivered to the Midwest Junto for the History of Science, April 24, 2003, manuscript in author's possession.

10. *National Defense Education Act*, Public Law 85-864, September 2, 1958, 72 *Stat.* 1580 20 U.S.C. § 401.

11. *Department of Defense Directive 5105.15, February 7, 1958*. It established the Advanced Research Projects Agency (ARPA). The ARPA Web site is the easiest place to find the directive. See http://www.darpa.mil/body/arpa_darpa.html.

12. The entire Kitchen Debate is preserved on "CNN Cold War—Historical Documents: Khrushchev-Nixon Debate," http://www.cnn.com/SPECIALS/cold.war/episodes/14/documents/debate/.

13. Criticisms recounted in Karal Ann Marling, *As Seen on TV: The Visual Culture of Everyday Life in the 1950s* (Cambridge, MA: Harvard University Press, 1998), pp. 278–80.

14. *Food Additives Amendment of 1958*, Public Law 85-929, September 6, 1958, 72 *Stat.* 1784.

15. Richard M. Nixon, "Acceptance Address, Delivered at the Republican National Convention, Chicago, Illinois, July 28, 1960," *Vital Speeches of the Day* 26 (August 15, 1960): 642–46.

16. John F. Kennedy, "Acceptance Address, Delivered at the Democratic National Convention, Los Angeles, California, July 15, 1960," *Vital Speeches of the Day* 26 (August 1, 1960): 610–12.

17. Dwight D. Eisenhower, "Liberty Is at Stake: A Farewell Address, Delivered to the Nation, Washington, DC, January 17, 1961," *Vital Speeches of the Day* 27 (February 1, 1961): 228–30.

18. Ibid.

19. Kennedy, "Acceptence," p. 611.

20. Jerome B. Wiesner, "Science in an Affluent Society," in *Where Science and Politics Meet*, ed. Jerome B. Wiesner (New York: McGraw-Hill, 1965), pp. 60–61.

21. John F. Kennedy, "Address at a White House Reception for Members of Congress and for the Diplomatic Corps of the Latin American Republics, March 13, 1961," *APPDA*.

22. *Clean Air Act*, Public Law 88-206, December 17, 1963, 77 *Stat.* 392.

23. John F. Kennedy, "Address at the Anniversary Convocation of the National Academy of Sciences, October 22, 1963," *APPDA*.

24. *An Act to Authorize an Institute of Child Health and Human Development, National Institutes of Health*, Public Law 87-838, September 16, 1962.

25. John F. Kennedy, "For the Freedom of Man, We Must All Work Together," delivered at inauguration ceremonies, Washington, DC, January 20, 1961, *Vital Speeches of the Day* 27 (February 1, 1961): 226–27.

26. Khrushchev's statements were reported in Robert S. McNamara, "United States Policy in Vietnam, 26 March 1964," Department of State Bulletin, April 13, 1964, p. 562; reprinted in *The Pentagon Papers*, Gravel ed., vol. 3, pp. 712–15.

27. John F. Kennedy, "Address at Rice University in Houston on the Nation's Space Effort, September 12, 1962," *APPDA*.

28. Ibid.

29. *Vocational Education Act of 1963*, Public Law 88-210.

Chapter 2

1. The definition is attributed to political columnists Rowland Evans and Robert Novak. It can be found at *American Experience: The Presidents*, http://www.pbs.org/wgbh/amex/presidents/36_l_johnson/l_johnson _domestic.html.

2. Lyndon B. Johnson, "Annual Message to the Congress on the State of the Union, January 8, 1964," *APPDA*.

3. Ibid.

4. Mark L. Gelfand, "The War on Poverty," in *The Johnson Years*, vol.1, ed. Robert A. Devine (Lawrence: University of Kansas Press, 1987), pp. 134–37.

5. Lyndon B. Johnson, "Remarks at the University of Michigan, May 22, 1964," *APPDA*.

6. United States Public Health Service, *Smoking and Health: Report of the Advisory Committee to the Surgeon General of the Public Health Service* (Washington, DC: GPO, 1964), p. 33.

7. Barry Goldwater, "Acceptance Speech, Delivered at the 1964 Republican National Convention, San Francisco, July 16, 1964," http://www.americanrhetoric.com/speeches/barrygoldwater1964rnc.htm.

8. The Daisy Girl commercial is housed at the Julian P. Kanter Political Commercial Archive, University of Oklahoma Political Communication Center.

9. Peter J. Kuznick, "Scientists on the Stump," *Bulletin of Atomic Scientists* 60 (November/December 2004): 32.

10. Ibid.

11. *Federal Cigarette Labeling and Advertising Act of 1965*, Public Law 89-92.

12. John Gardner, *Report of the President's Task Force on Education, November 14, 1964*, p. 34. The report is available at the LBJ Presidential Library, Austin, Texas.

13. *Higher Education Act of 1965*, Public Law 89-329, *Stat.* 79 (1965):1219.

14. *A National Program to Conquer Heart Disease, Cancer and Stroke, Report to the President, the President's Commission on Heart Disease, Cancer and Stroke* (Washington, DC: GPO, 1964), pp. 56–73.

15. *Heart Disease, Cancer, and Stroke Amendments of 1965*, Public Law 89-239.

16. See, for example, Don K. Price, *The Scientific Estate* (Cambridge, MA: Belknap Press of Harvard University Press, 1965); Donald W. Cox, *America's*

New Policy Makers; the Scientists' Rise to Power (Philadelphia: Chilton Books, 1964); Ralph E. Lapp, *The New Priesthood; the Scientific Elite and the Uses of Power* (New York: Harper & Row, 1965); and Spencer Klaw, *The New Brahmins; Scientific Life in America* (New York: Morrow, 1968). The bricklayer quote can be found in Emmanuel G. Mesthene, "Can Only Scientists Make Government Science Policy?" *Science* 145 (July 17, 1964): 237.

17. P. H. A., "Plans for *Science*," *Science* 138 (October 19, 1962): 405.

18. Warren Weaver's comments about science appeared in Paul Weiss, "Bringing the Public to Science," *Vital Speeches of the Day* 32 (January 15, 1966): 223.

19. Ken Hechler, ed., *Toward the Endless Frontier: History of the Committee on Science and Technology, 1959–79* (Washington, DC: GPO, 1980), p. 238.

20. Ibid., p. 276.

21. Philip H. Abelson is cited in Daniel Kevles, *The Physicists: The History of a Scientific Community in Modern America* (New York: Vintage, 1979), p. 417.

22. For the younger generation as *Time*'s Man of the Year, see "The Inheritor," *Time*, January 6, 1967, at http://www.time.com/time/magazine/archives.

23. Quotations are from Philip H. Abelson, "A Partisan Attack on Research," *Science* 156 (June 9, 1967): 1315; and J. V. Reistrup, "The Moral Sense of Scientists," *Science* 155 (January 20, 1967): 271.

24. D. S. Greenberg, "Money for Science: The Community Is Beginning to Hurt," *Science* 152 (June 10, 1966): 1485.

25. Michael Hauben, "Behind the Net—The Untold History of the ARPANET," http://www2.dei.isep.ipp.pt/docs/arpa—1.html.

26. D. S. G., "Scientists in Politics," *Science* 161 (July 12, 1968): 145–46.

27. Richard Nixon, "Accepting the Republican Nomination for President, Miami, August 8, 1968," http://www.theamericanpresidency.us/nomanixon68.htm.

28. John Walsh, "Richard M. Nixon: Promises of a Shift in Priorities," *Science* 162 (October 18, 1968): 335–37.

29. Philip B. Boffey, "Humphrey vs. Nixon: Candidates Sharpen the Science Issues," *Science* 162 (November 1, 1968): 549–51.

30. Philip B. Boffey, "Scientists in Politics: Humphrey's Group Outshines Nixon's," *Science* 162 (October 11, 1968): 244.

31. Emilio Q. Daddario, "Academic Science and the Federal Government," *Science* 162 (December 13, 1968): 1249–51; and Don K. Price, "Purists and Politicians," *Science* 163 (January 3, 1969): 25–31.

32. Bryce Nelson, "MIT's March 4: Scientists Discuss Renouncing Military Research," *Science* 163 (March 14, 1969): 1175–78.

33. James K. Glassman, "AAAS Boston Meeting: Dissenters Find a Forum," *Science* 167 (January 2, 1970): 36–38.

34. Peter Thomson, "TRACES: Basic Research Links to Technology Appraised," *Science* 163 (January 24, 1969): 374–75; Peter Thompson, "NEH Urges Annual 'Social Report,'" *Science* 163 (January 31, 1969): 456; Donald F. Hornig, "United States Science Policy: Its Health and Future Direction," *Science* 163 (February 7, 1969): 525–28; and James A. Shannon, "Science and Social Purpose," *Science* 163 (February 21, 1969): 769–73.

35. Eugene B. Skolnikoff, "Public Challenge of Government Action," *Science* 164 (May 2, 1969): 499.

36. Bryce Nelson, "ABM: Scientists Are Important in Building Senate Opposition," *Science* 164 (May 9, 1969): 654–56.

37. D. S. Greenberg, "Science under Nixon: Influence Has Declined in National Affairs," *Science* 169 (September 11, 1970): 1056–57.

38. Bryce Nelson, "NSF Directorship: Why Did Nixon Veto Franklin A. Long?" *Science* 164 (April 25, 1969): 406–11.

39. Ibid.; P. M. B., "Nixon Chooses OST Deputy Director," *Science* 164 (June 13, 1969): 1263.

40. Lee A. DuBridge, "Science Serves Society," *Science* 164 (June 6, 1969): 1137–40.

41. Deborah Shapley, "Scientist and Public: Chapter and Verse from David," *Science* 172 (June 13, 1971): 1010.

42. *National Environmental Policy Act of 1969*, Public Law 91-190, 42 U.S.C. 4321–4347, January 1, 1970.

43. Ibid.

44. Richard M. Nixon, "Special Message to the Congress about Reorganization Plans to Establish the Environmental Protection Agency and the National Oceanic and Atmospheric Administration, July 9, 1970," *APPDA*.

45. Ken Hechler, ed., *Toward the Endless Frontier: History of the Committee on Science and Technology, 1959–79* (Washington, DC: GPO, 1980), pp. 149–61.

46. James H. Krieger, "Technology Assessment No Longer Theoretical," *Chemical and Engineering News* 243 (April 6, 1970): 65–68.

47. B. J. C., "Scientists and the Public Interest," *Science* 182 (October 26, 1973): 367.

48. For the DES story and the quotations, see Alan I Marcus, *Cancer from Beef: DES, Federal Food Regulation, and Consumer Confidence* (Baltimore: Johns Hopkins University Press, 1993).

49. D. S. Greenberg, "Research Priorities: New Program at NSF Reflects Shift in Values," *Science* 170 (October 9, 1970): 144–46.

50. Richard M. Nixon, "Special Message to the Congress on Science and Technology, March 16, 1972," *APPDA*.

51. Richard M. Nixon, "Message to the Congress Transmitting Reorganization Plan 1 of 1973 Restructuring the Executive Office of the President, January 26, 1973," *APPDA*.

52. Philip H. Abelson, "Additional Sources of Financial and Political Support of Science," *Science* 180 (April 20, 1973): 259; and John Walsh, "Killian Committee: Report Urges Advisory Council in White House," *Science* 185 (July 5, 1974): 39–41.

53. Robert Gillette, "Advising the White House: NSF Says the New System Works," *Science* 185 (July 26, 1974): 334–36; and "Science Advisor Stever: The View from 1800 G Street," *Science* 185 (July 26, 1974): 335.

54. Richard M. Nixon, "Remarks on Presenting the National Medal of Science Awards for 1973, October 10, 1973," *APPDA*.

55. Gerald R. Ford, "Address before a Joint Session of the Congress Reporting on the State of the Union, January 15, 1975," *APPDA*.

56. William D. Carey, "The Ford Budget: New Signals for Science," *Science* 187 (February 28, 1975): 705.

57. Philip M. Boffey, "Energy Research: A Harsh Critique Says Federal Effort May Backfire," *Science* 190 (November 7, 1975): 535–37.

58. Deborah Shapley, "Congress: House Votes Veto Power on All NSF Research Grants," *Science* 188 (April 25, 1975): 338.

59. Philip M. Boffey, "Nuclear Safety: A Federal Adviser's Warnings Provoke Ire of Colleagues," *Science* 192 (June 4, 1976): 978–79.

60. Nicholas Wade, "Recombinant DNA: Guidelines Debated at Public Hearing," *Science* 191 (February 27, 1976): 834–36.

61. Nicholas Wade, "Recombinant DNA: NIH Group Stirs Storm by Drafting Laxer Rules," *Science* 190 (November 21, 1975): 767–69; and Nicholas Wade, "Recombinant DNA: NIH Sets Strict Rules to Launch New Technology," *Science* 190 (December 19, 1975): 1175–79.

62. Joel R. Primack and Frank von Hippel, *Advice and Dissent: Scientists in the Public Arena* (New York: New American Library, 1976).

63. Barbara J. Culliton, "Kennedy: Pushing for More Public Input in Research," *Science* 188 (June 20, 1975): 1187–89.

64. US Senate, *Report on the National Policy and Priorities for Science and Technology Act of 1974*, #93-1254, October 9, 1974.

65. Hechler, *Toward the Endless Frontier*, pp. 612–25. Also see *US*

Statutes, "Legislative History of P. L. 94-282 Science, Technology and Priorities," pp. 885–95.

CHAPTER 3

1. Luther J. Carter, "Jimmy Carter's Advisors: Drawing from the Public Interest Movement," *Science* 193 (September 3, 1976): 868–70.

2. Barbara J. Culliton, "Stringent New Ethics Law Worries Government Scientists," *Science* 203 (March 9, 1979): 981.

3. The quotes are from Alan I Marcus, "Sweets for the Sweet: Saccharin, Knowledge and the Contemporary Regulatory Nexus," *Journal of Policy History* (Winter 1996): 33–47. For Kennedy, see Nicholas Wade, "Kennedy Leaves as FDA Commissioner," *Science* 203 (July 13, 1979): 173.

4. Jimmy Carter, "National Energy Plan—Address Delivered before a Joint Session of the Congress, April 20, 1977," *APPDA.*

5. Luther J. Carter, "Carter Energy Message: How Stiff a Prescription?" *Science* 196 (May 6, 1977): 630–32.

6. Jimmy Carter, "Solar Energy Message to the Congress, June 20, 1979," *APPDA.*

7. Jimmy Carter, "Economic Renewal Program Remarks Announcing the Program, August 28, 1980," *APPDA;* and Robert Reinhold, "Carter Asks Action to Spur Technology," *New York Times,* November 1, 1979, p. A1.

8. Robert Reinhold, "Information Banks Abstracts," *New York Times,* November 1, 1979, p. 1.

9. Luther J. Carter, "Industrial Productivity and the 'Soft Sciences,'" *Science* 209 (July 25, 1980): 476–77.

10. John Walsh, "NSF under Challenge from Congress, Engineers," *Science* 209 (September 26, 1980): 1499.

11. Stuart Eisenstat's quote can be found in Barry Krusch, "Why We Need a New Constitution," http://www.totse.com/en/politics/political_spew/newconst.html.

12. Reagan campaign rhetoric can be found in Ronald Reagan, "Acceptance of the Republican Nomination for President, July 17, 1980," http://www.medaloffreedom.com/RonaldReaganAcceptance.htm; and Ronald Reagan, "Inaugural Address, January 20, 1981," *APPDA.*

13. William D. Carey, "Affordable Science," *Science* 212 (May 1, 1981): 497.

14. Colin Norman, "Science Advisor Post Has Nominee in View," *Science* 212 (May 22, 1981): 903–904.

15. Robert Reinhold, "Physicist Is Named as Science Advisor," *New York Times*, May 20, 1981, p. A32; Philip J. Hilts, "Science Advisor Urges Focusing Research in Productive Fields," *Washington Post*, June 26, 1981, p. A14; Philip J. Hilts, "Physicist, Relative Unknown, Named as Reagan's Science Advisor," *Washington Post*, June 14, 1981, p. A12; and Barbara J. Culliton, "Keyworth Gives First Policy Speech," *Science* 213 (July 10, 1981): 183–84.

16. R. Jeffrey Smith, "Reagan Proposes Huge Nuclear Buildup," *Science* 214 (October 16, 1981): 309.

17. Robert Reinhold, "U.S. Science Agency's New Chief Pushes for Change," *New York Times*, September 29, 1981, p. C1; Robert Reinhold, "Reagan Aides Fail to Mollify Worried Scientists," *New York Times*, October 27, 1981, p. C3; Christine Russell, "Scientists Fear Budget Cutbacks," *Washington Post*, October 27, 1981, p. A7; Christine Russell, "Science Leaders Fear 'Irreversible Damage' from Budget Cuts," *Washington Post*, October 28, 1981, p. A15; Robert Reinhold, "U.S. Science Aid Study Asked," *New York Times*, October 28, 1981, p. A24; and Arlen J. Large, "Science Elite Urges Reagan to Shun Plan for Cuts in Research," *Wall Street Journal*, October 28, 1981, p. 42.

18. Robert C. Cowen, "Our Present System of Science/Math Education Is Letting the US Down," *Christian Science Monitor*, June 9, 1982, p. 16; Robert C. Cowen, "Concern for US Technology Lag May Spur R&D Renaissance," *Christian Science Monitor*, October 22, 1982, p. 5; and Jonathan Harsch, "Knock Down Walls, Says Reagan Science Advisor," *Christian Science Monitor*, October 29, 1982, p. B4.

19. Colin Norman, "Keyworth Says Cuts May Be Good for Science," *Science* 215 (January 1, 1982): 39; and Arlen J. Large, "Science Panel Says U.S. Shouldn't Classify Research Unless the Soviets Would Benefit," *Wall Street Journal*, October 1, 1982, p. 37.

20. Ronald Reagan, "Message to the Congress Transmitting the Annual Science and Technology Report, April 21, 1982," *APPDA*.

21. James Gleick, "Reagan, Citing Foreign Challenge, Outlines Superconductivity Plan," *New York Times*, July 29, 1987, p. A1.

22. John Walsh, "NSF Lends a Hand with DOD Award," *Science* 238 (November 16, 1987): 748; and M. Mitchell Waldrip, "NSF Commits to Supercomputers," *Science* 228 (May 3, 1985): 568–71.

23. Ronald Reagan, "Executive Order 12291—Federal Regulation, February 17, 1981," *APPDA*.

24. William D. Ruckelshaus, "Science, Risk, and Public Policy," *Science* 221 (September 9, 1983): 1026–28.

25. Barbara J. Culliton, "Watson Fights Back," *Science* 228 (April 12, 1985): 160.

26. Peter Behr, "Schultze Attacks Premise of New Industrial Policy," *Washington Post*, September 29, 1983, p. B1.

27. Peter Behr, "Panel on Competitiveness Urges Spending Cuts," *Washington Post*, December 7, 1984, p. F4.

28. Calvin Sims, "Business-Campus Ventures Grow," *New York Times*, December 14, 1987, p. D1.

29. Alan L. Otten, "Coalition Lobbies for Applied and Basic Research, Rallying behind the Banner of Competitiveness," *Wall Street Journal*, March 3, 1987, p. 70.

30. Philip M. Boffey, "White House Backs New Mode for Key Research," *New York Times*, February 1, 1987, p. 30.

31. Ronald Reagan, "Executive Order 12591—Facilitating Access to Science and Technology, April 10, 1987," *APPDA*.

32. *Malcolm Baldrige National Quality Improvement Act of 1987*, Public Law 100-107.

33. Larry Thompson, "The Research Budget," *Washington Post*, February 23, 1988, p. HE9.

34. Philip M. Boffey, "Two Leaders Challenge the 'Big Science' Trend," *New York Times*, May 3, 1988, p. C1.

35. George H. W. Bush and Michael S. Dukakis, "Science Policy," *Science* 242 (October 14, 1988): 173–78; and "Politics and the Laboratory," *Washington Post*, November 6, 1988, p. C3.

Chapter 4

1. Colin Norman, "Inaction on Technology Programs Stirs Congress," *Science* 244 (April 14, 1989): 137–38.

2. Bob Stone, *Confessions of a Civil Servant: Lessons in Changing America's Government and Military* (Lanham, MD: Rowman & Littlefield, 2003), pp. 42–46; Alan I Marcus and Howard P. Segal, *Technology in America: A Brief History*, 2nd ed. (Belmont, CA: Thomson, 1999), pp. 339–40.

3. Colin Norman, "Dear President-Elect . . . : A Place at the Head Table," *Science* 243 (January 13, 1989): 163.

4. William J. Broad, "Panel Rejects Fusion Claim, Urging No Federal Spending," *New York Times*, July 13, 1989, p. A1; and "U.S. Panel Finds No Evidence of Cold Fusion," *New York Times*, November 4, 1989, p. 9.

5. Mark Crawford, "Senate Committee Quizzes Bromley," *Science* 245 (July 28, 1989): 349.

6. George H. W. Bush, "Remarks to the National Academy of Sciences, April 23, 1990," *APPDA.*

7. William Booth, "Scientists Say Basic Research Is Hurt by Cut in New Grants," *Washington Post*, April 3, 1990, p. A17; Robert C. Cowen, "U.S. Science Budget Set to Increase," *Christian Science Monitor*, April 30, 1990, p. 8; and William J. Broad, "Big Science: Is It Worth the Price?" *New York Times*, May 27, 1990, sec. 1, part 1, p. 1.

8. Ibid.

9. Leon Lederman, "Science: The End of the Frontier," supplement to *Science* 251(January 11, 1991).

10. William Booth, "Science and the Art of Money," *Washington Post*, February 17, 1991, p. C1.

11. Malcolm W. Browne, "Budget Threatens Physics Project," *New York Times*, May 19, 1991, sec. 1, p. 19.

12. Joseph Palca, "OTA Challenges Dogma on Research Funding," *Science* 251 (March 29, 1991): 1555.

13. Michael Schrage, "Industrial Policy by Another Name," *Washington Post*, February 15, 1991, p. F3.

14. Evelyn Richards, "White House Lists 22 Key Technologies," *Washington Post*, April 26, 1991, p. F1.

15. Martin Tolchin, "House Plan Fuels Debate on Business," *New York Times*, July 22, 1991, p. D3.

16. Joseph Palca, "Proposals to Limit Indirect Costs Emerge in Congress," *Science* 252 (May 17, 1991): 910.

17. Eliot Marshall, "Bush Tries Trimming R&D Pork," *Science* 255 (March 27, 1992): 1635.

18. Richard Stone, "Peer Review Catches Congressional Flak," *Science* 256 (May 15, 1992): 959.

19. J. P., "Congress Sends a Message," *Science* 256 (May 29, 1992): 1274; and Richard S. Nicholson, "Congressional Pork versus Peer Review," *Science* 256 (June 11, 1992): 1497.

20. Joseph Palca, "Congress Queries Hallowed Principles," *Science* 257 (September 18, 1992): 1620.

21. Bill Clinton, "Technology for America's Economic Growth, February 22, 1993," http://simr02.si.ehu.es/DOCS/nearnet.gnn.com/mag/10_93/articles/clinton/clinton.tech.html.

22. Ibid.

23. David P. Hamilton, "Who's Who among Science Advisors," *Science* 258 (October 16, 1992): 393.

24. Daniel S. Greenberg, "Gore May Become Research Czar," *Cleveland Plain Dealer*, November 16, 1992, p. 5B.

25. Christopher Anderson, "Clinton Picks His Science Advisor," *Science* 259 (January 8, 1993): 171.

26. Robert C. Cowen, "US Science Magnifies Its Social Role," *Christian Science Monitor*, February 10, 1993, p. 9; and Daniel E. Koshland Jr., "Basic Research (1)," *Science* 259 (January 15, 1993): 291.

27. Thomas W. Lippman, "In Senate, Committee Turf Race Is on for Clinton's Technological Initiative," *Washington Post*, March 3, 1993, p. A15.

28. John Kamensky, "National Partnership for Reinventing Government (Formerly the National Performance Review): A Brief History, January 1999," http://www.npr.gov/whoweare/history2.html.

29. Jeffrey Mervis, "Clinton Moves to Manage Science," *Science* 261 (September 24, 1993): 1668–69.

30. "Clinton Vows to Be Ally in Breast Cancer Fight," *St. Louis Post-Dispatch*, October 19, 1993, p. 4B.

31. Daniel E. Koshland Jr., "Moderation in Science Budgeting," *Science* 260 (April 30, 1993): 603.

32. "Scientists Urge Coordination of Research Groups," *Atlanta Journal-Constitution*, August 14, 1993, p. E1.

33. Alex Barnum, "White House to Get Science Clout," *San Francisco Chronicle*, October 27, 1993, p. D1.

34. Jeffrey Mervis, "OSTP Plans a Blueprint for Research," *Science* 263 (January 14, 1994): 165–66.

35. Bill Clinton, *Science in the National Interest* (Washington, DC: Office of Science and Technology Policy, 1994).

36. Boyce Rensberger, "Science Policy Is Welcomed Despite 'Highly Unrealistic' Hopes for More Funding," *New York Times*, August 5, 1994, p. A15.

37. Michael Schrage, "Look for the New Makeup in Congress to Shake Up High-tech Partnerships," *Washington Post*, November 11, 1994, p. C3.

38. Andrew Lawler, "Science and Technology Policy Headed for Political Maelstrom," *Science* 266 (November 18, 1994): 1152–53.

39. Bruce Babbitt, "Science: Opening the Next Chapter of Conservation History," *Science* 267 (March 31, 1995): 1954–55.

40. Andrew Lawler, "GOP Plans Would Reshuffle Science," *Science* 268 (May 19, 1995): 964–65, 967; and Richard Stone, "Senate May Strengthen Science in Risk Bill," *Science* 269 (July 14, 1995): 151.

41. Andrew Lawler, "Democrats Urge Clinton to Stand Up for R&D Programs," *Science* 269 (September 29, 1995): 1810; and Andrew Lawler, "Clinton Defends R&D in Partisan Speech," *Science* 270 (October 27, 1995): 571–72.

42. Andrew Lawler, "Research Knows No Season as Budget Cycle Goes Awry," *Science* 271 (February 2, 1996): 589.

43. Lee Bowman, "Gore Calls GOP Approach to Science Flintstonian," *Denver Rocky Mountain News*, February 13, 1996, p. 22A.

44. Andrew Lawler, "Clinton's R&D Achievements Tilt toward Technology," *Science* 271 (February 23,1996): 1049–50.

45. Andrew Lawler, "Legislators Get into the Details," *Science* 273 (July 5, 1996): 25–26; and Andrew Lawler, "Agency Heads See Give in R&D Plan," *Science* 273 (August 2, 1996): 572.

46. Daniel S. Greenberg, "Clinton: False Friend of Science," *Journal of Commerce*, October 21, 1996, p. 9A.

47. Andrew Lawler, "Panels Lead the Way on the Road to Kyoto Conference," *Science* 278 (October 10, 1997): 216–17.

48. Bill Clinton, "Address before a Joint Session of the Congress on the State of the Union, January 27, 1998," *APPDA*.

49. Ibid.

50. Warren E. Leary, "Clinton Seeks $170 Billion for Research in Budget," *New York Times*, February 3, 1998, p. F4.

51. Eliot Marshall, "U.S. R&D Budget Becomes Political Football," *Science* 281 (July 3, 1998): 16–17.

52. Bill Clinton, "Executive Order 13100—President's Council on Food Safety, August 25, 1998," *APPDA*.

53. Nancy McVicar, "Cloning-related Work Raises Flags," *New Orleans Times-Picayune*, November 19, 1998, p. A17.

54. Tim Weiner, "Colleges Master Congress," *New Orleans Times-Picayune*, August 24, 1999, p. A1.

55. David Malakoff, "Balancing the Science Budget," *Science* 287 (February 11, 2000): 952–55; and Myles Brand, "Wise Nation Puts Money in Knowledge," *Chicago Sun-Times*, March 5, 2000, p. 38.

56. David Malakoff, "2001 U.S. Budget: Record Year for Science, but Can It Be Repeated?" *Science* 291 (January 5, 2001): 33.

57. Jeffrey Mervis, "National Science Foundation: Transition Rumor Targets Colwell," *Science* 291 (January 26, 2001): 572.

58. Dan Vergano, "Medical Research Has Healthy Budget," *USA Today*, March 20, 2001, p. 9D.

59. Carolyn Said, "VC to Lead Efforts to Shape Tech Policy," *San Francisco Chronicle*, March 29, 2001, p. B1; and David Malakoff, "Bush Appointment," *Science* 292 (April 6, 2001): 28–29.

60. Donald Kennedy, "An Unfortunate U-Turn on Carbon," *Science* 291 (March 30, 2001): 2515.

61. D. Allan Bromley, "Science and Surpluses," *New York Times*, March 9, 2001, p. A19.

62. David Malakoff, "U.S. Congress: Now Batting for Science: New York's Sherry Boehlert," *Science* 292 (May 11, 2001): 1048–49.

63. Andrew Lawler, "White House: President's New Advisor Ready to Put Science in Its Place," *Science* (June 29, 2001): 2408–409.

64. Diana Jean Schemo, "Sept. 11 Strikes at Labs' Doors," *New York Times*, August 13, 2002, p. F1.

65. Mary Leonard, "Fighting Terror Planned Restrictions on Campus," *Boston Globe*, October 11, 2002, p. A21.

66. *Federal Advisory Committee Act*, Public Law 92-463, *Stat.* 86 (October 6, 1972): 770.

67. David Michaels, "Advice without Dissent," *Science* 298 (October 25, 2002): 703.

68. Glen Martin, "Agriculture Secretary Pushes New Crops," *San Francisco Chronicle*, June 24, 2003, p. A3.

69. "U.S. National Academies Move to Review Advisory Panels," *Science* 299 (February 28, 2003): 1295.

70. House Committee on Government Reform—Minority Staff Special Investigations Division, *Politics and Science in the Bush Administration*, August 2003, prepared for Rep. Henry A. Waxman (updated November 13, 2003), http://www.reform.house.gov/min.

71. Roger Pielke Jr., "Another Epidemic of Politics," *Science* 300 (May 16, 2003): 1092–93.

72. Andrew C. Revkin, "White House Proposes Reviews for Studies on New Regulations," *New York Times*, August 29, 2003, p. A12.

73. Michael Slackman, "Gore Environmental Speech Becomes an Assault on Bush," *New York Times*, January 16, 2004, p. A16; and "Al Gore Speaks on Global Warming and the Environment, Beacon Theater, New York, January 15, 2004, Noon," http://civic.moveon.org/gore3/speech.html.

74. Dan Vergano, "White House Manipulates Science, Leaders in Field Say," *USA Today*, February 19, 2004, p. 10A.

75. Emily Johns, "Scientists Criticize White House," *Minneapolis Star*

Tribune, February 19, 2004, p. 8A; James Glanz, "Scientists Say Administration Distorts Facts," *New York Times*, February 19, 2004, p. A18; Carl T. Hall, "Top Scientists Accuse Bush of Manipulating Research for Political Gain," *San Francisco Chronicle*, February 19, 2004, p. A7; and James Glanz, "At the Center of the Storm over Bush and Science," *New York Times*, March 30, 2004, p. F1.

76. David Malakoff, "White House Rebuts Charges It Has Politicized Science," *Science* 304 (April 9, 2004): 184–85.

77. William J. Broad, "Science Group Says U.S. Budget Plan Would Harm Research," *New York Times*, April 23, 2004, p. A20; and William J. Broad, "U.S. Is Losing Its Dominance in the Sciences," *New York Times*, May 3, 2004, p. A1.

78. "Kerry Accuses Bush of Politicizing Science," *Denver Post*, June 22, 2004, p. B-06.

79. Jeff Nesmith, "Politics Is Intruding on Science, Panel Told," *Atlanta Journal-Constitution*, July 22, 2004, p. 5A; and Jeffrey Mervis, "Congressmen Clash on Politics and Scientific Advisory Committees," *Science* 305 (July 30, 2004): 593.

80. Kenneth Chang, "Scientists Begin a Campaign to Oppose President's Policies," *New York Times*, September 28, 2004, p. A17.

81. Donald Kennedy, "The Candidates Speak," *Science* 306 (October 1, 2004): 19.

82. Jeffrey Mervis, "Bush Victory Leaves Scars—and Concerns about Funding," *Science* 306 (November 12, 2004): 1110–13.

83. Ibid.

84. Alan I. Leshner, "A Dangerous Signal to Science," *Science* 306 (December 24, 2004): 2163.

85. Jeffrey Mervis, "Caught in the Squeeze," *Science* 307 (February 11, 2005): 832.

86. Ibid.

87. Jeffrey Mervis, "Marburger Asks Social Scientists for a Helping Hand in Interpreting Data," *Science* 308 (April 29, 2005): 617.

INDEX